Using Stable Diffusion with Python

Leverage Python to control and automate high-quality AI image generation using Stable Diffusion

Andrew Zhu (Shudong Zhu)

Using Stable Diffusion with Python

Group Product Manager: Niranjan Naikwadi
Publishing Product Manager: Sanjana Gupta
Senior Editor: Tazeen Shaikh
Senior Content Development Editor: Joseph Sunil
Technical Editor: Rahul Limbachiya
Copy Editor: Safis Editing
Project Coordinator: Shambhavi Mishra
Proofreader: Tazeen Shaikh
Indexer: Tejal Daruwale Soni
Production Designer: Alishon Mendonca
Marketing Coordinator: Vinishka Kalra

First published: June 2024

Production reference: 1170524

Published by Packt Publishing Ltd.
Grosvenor House
11 St Paul's Square
Birmingham
B3 1RB, UK

ISBN 978-1-83508-637-7

www.packtpub.com

To my beloved wife, Yinhua Fan, and our precious sons, Charles Zhu and Daniel Zhu.

You are the spark that ignites my creativity, the fuel that drives my passion, and the love that sustains me throughout this journey. Without your unwavering support, encouragement, and inspiration, this book would not have been possible.

–Andrew Zhu

Foreword

Artificial intelligence has ushered in a new era of creativity, with generative models offering a glimpse into what was once futuristic. Stable Diffusion stands out as an innovative leap forward, blending technical sophistication with practical application in a way that empowers creators across diverse domains.

Andrew Zhu's book is a comprehensive resource for understanding Stable Diffusion's technical underpinnings. He provides an in-depth exploration of its foundational principles, contrasts it with alternative generative models, and demonstrates how to apply it to varied creative fields.

Technologies like Stable Diffusion are catalysts for creativity, offering accelerated workflows for content refinement, editing, and generation. They harness cutting-edge optimization techniques to produce detailed and unique imagery, setting a new standard in the efficiency of image creation and enabling high-quality results that were once only attainable through meticulous manual effort. Whether you're building data-driven applications or experimenting with imaginative visual storytelling, the model's robust and scalable architecture is designed to seamlessly integrate into diverse production environments.

The pace of AI advancements, especially in deep learning, is unparalleled. We're witnessing unprecedented growth as new architectures and optimization techniques continuously redefine the state-of-the-art. Stable Diffusion stands as a testament to this rapid evolution, providing practical solutions that enable independent creators and organizations alike to quickly translate their ideas into reality, often reducing what once took weeks to mere hours.

These are thrilling times to be exploring Stable Diffusion. Andrew's carefully curated insights will undoubtedly inspire innovative projects that harness the true power of generative AI. Dive into this book with curiosity and enthusiasm, and you'll embark on a journey that will not only inform but ignite new creative possibilities.

Andrew has done groundbreaking work in understanding the nuances of generative models and creating practical methodologies that push the boundaries of what's possible. His book stands as a testament to this ingenuity, providing insights that will guide both enthusiasts and professionals toward the next era of creativity and technological progress.

Enjoy the read! The adventure is just beginning.

-Matt Fisher, Co-Founder and CTO, Dahlia Labs

Contributors

About the author

Andrew Zhu (Shudong Zhu) is a seasoned Microsoft applied data scientist with over 15 years of experience in the tech industry. Renowned for his exceptional ability to distill complex machine learning and AI concepts into engaging, informative narratives, Andrew regularly contributes to esteemed publications such as Toward Data Science. His previous book, *Microsoft Workflow Foundation 4.0 Cookbook*, earned a commendable 4.5-star average rating on Amazon.

As a contributor to the popular Hugging Face Diffusers library, a leading Stable Diffusion Python library and a primary focus of this book, Andrew brings unparalleled expertise to the table. Currently, he leads the AI department at a stealth start-up company, leveraging his extensive research background and proficiency in generative AI to transform the online shopping experience and pioneer the future of AI in retail.

Outside of his professional pursuits, Andrew resides in Washington State, USA, with his loving family, including his two sons.

About the reviewers

Krishnan Raghavan is an IT professional with over 20 years of experience in the areas of software development and delivery excellence across multiple domains and technologies, ranging from C++ to Java, Python, Angular, Golang, and data warehousing.

When not working, Krishnan likes to spend time with his wife and daughter, as well as reading fiction, nonfiction, and technical books and participating in hackathons. Krishnan tries to give back to the community; he is a part of the GDG – Pune volunteer group, helping the team to organize events.

You can connect with Krishnan at mailtokrishnan@gmail.com.

I would like to thank my wife, Anita, and daughter, Ananya, for giving me the time and space to review this book.

Swagata Ashwani is a data professional with over seven years of experience in the healthcare, retail, and platform integration industries. She is an avid blogger and writes about state-of-the-art developments in the AI space. She is particularly interested in **Natural Language Processing (NLP)** and focuses on researching how to make NLP models work in a practical setting. She is also a podcast host and chapter lead of Women in Data, and loves to advocate for women in technology roles and the responsible use of AI in this fast-paced era. In her spare time, she loves to play her guitar, sip masala chai, and find new spots for doing yoga.

Table of Contents

6

Using Stable Diffusion Models   65

Part 2 – Improving Diffusers with Custom Features

7

Optimizing Performance and VRAM Usage   79

8

Using Community-Shared LoRAs   89

9

Using Textual Inversion   107

10

Overcoming 77-Token Limitations and
Enabling Prompt Weighting   119

11

Image Restore and Super-Resolution   141

Part 3 – Advanced Topics

13

14

15

Part 4 – Building Stable Diffusion into an Application

18

Applications – Object Editing and Style Transferring   251

19

Generation Data Persistence   263

20

Creating Interactive User Interfaces   269

21

22

Preface

When Stable Diffusion was released on August 22, 2022, this Diffusion-based image generation model quickly caught the attention of the whole world. Both its model and source code are completely open source and hosted on GitHub. With millions of community participants and users, numerous new and mixed models have been released. Tools such as Stable Diffusion WebUI and InvokeAI have been created.

While the Stable Diffusion WebUI tool can generate fantastic images driven by the diffusion model, its usability is limited. The open source Diffusers package from Hugging Face allows users to have full control over Stable Diffusion using Python. However, it lacks many key features, such as loading custom LoRA models and Textual Inversion, utilizing community-shared models/checkpoints, scheduling and weighted prompts, unlimited prompt tokens, fixing the resolution of the images, and upscaling. This book will assist you in overcoming the limitations of Diffusers and implementing the advanced features to create a fully customized and industrial-level Stable Diffusion application.

By the end of this book, you will be able to not only use Python to generate and edit images but also leverage the solutions provided in the book to build Stable Diffusion applications for your business and users.

Who this book is for

This book is for AI image and art generation enthusiasts who want to gain a comprehensive understanding of how image generation and diffusion models work.

It is also for artists who are interested in fully understanding AI image generation and gaining precise control over the Diffusion model.

Python application developers who aim to create AI image generation applications based on Stable Diffusion will also find this book helpful.

Finally, this book is aimed at data scientists, machine learning engineers, and researchers who want to programmatically control the Stable Diffusion process, automate pipelines, build custom pipelines, and conduct testing and validation using Python.

What this book covers

Chapter 1, Introducing Stable Diffusion, provides an introduction to the AI image generation technology Stable Diffusion.

Chapter 2, Setting Up an Environment for Stable Diffusion, covers how to set up the CUDA and Python environments to run Stable Diffusion models.

Chapter 3, Generate Images Using Stable Diffusion, is a whirlwind chapter that helps you start using Python to generate images with Stable Diffusion.

Chapter 4, Understand the Theory behind the Diffusion Models, digs into the internals of the Diffusion model.

Chapter 5, Understanding How Stable Diffusion Works, covers the theory behind Stable Diffusion.

Chapter 6, Using the Stable Diffusion Model, covers model data handling and converting and loading model files.

Chapter 7, Optimizing Performance and VRAM Usage, teaches you how to improve performance and reduce VRAM usage.

Chapter 8, Using Community-Shared LoRAs, shows you how to use the community-shared LoRAs with Stable Diffusion checkpoint models.

Chapter 9, Using Textual Inversion, uses the community-shared Textual Inversion with Stable Diffusion checkpoint models.

Chapter 10, Unlocking 77 Token Limitations and Enabling Prompt Weighting, covers how to build custom prompt-handling code to use unlimited-size prompts with weighted importance scores. Specifically, we'll explore how to assign different weights to individual prompts or tokens, allowing us to fine-tune the model's attention and generate more accurate results.

Chapter 11, Image Restore and Super-Resolution, shows you how to fix and upscale images using Stable Diffusion.

Chapter 12, Scheduled Prompt Parsing, shows you how to build a custom pipeline to support scheduled prompts.

Chapter 13, Generating Images with ControlNet, covers how to use ControlNet with Stable Diffusion checkpoint models.

Chapter 14, Generating Video Using Stable Diffusion, shows you how to use AnimateDiff together with Stable Diffusion to generate a short video clip and understand the theory behind the video generation.

Chapter 15, Generating Image Descriptions Using BLIP-2 and LLaVA, covers how to use large language models (LLMs) to extract descriptions from images.

Chapter 16, Exploring Stable Diffusion XL, shows you how to start using Stable Diffusion XL, a newer and better Stable Diffusion model.

Chapter 17, Building Optimized Prompts for Stable Diffusion, discusses techniques to write up Stable Diffusion prompts to generate better images, as well as leveraging LLMs to help generate prompts automatically.

Chapter 18, Applications: Object Editing and Style Transferring, covers how to use Stable Diffusion and related machine learning models to edit images and transfer styles from one image to another.

Chapter 19, Generation Data Persistence, shows you how to save image generation prompts and parameters into the generated PNG image.

Chapter 20, Creating Interactive User Interfaces, shows you how to build a Stable Diffusion WebUI using the open source framework Gradio.

Chapter 21, Diffusion Model Transfer Learning, covers how to train a Stable Diffusion LoRA from scratch.

Chapter 22, Exploring Beyond Stable Diffusion, provides additional information about Stable Diffusion, AI, and how to learn more about the latest developments.

To get the most out of this book

You will need to have some experience with the Python programming language. Familiarity with neural networks and PyTorch will be helpful for reading and running up the code in this book.

Disclaimer:
This book has been written keeping in mind ethical practices and regulations. Please avoid utilizing the knowledge that you will get here for any unethical purposes. Please refer to Chapter 22 for an in-depth review of the ethics of using AI.

Software/hardware covered in the book	Operating system requirements
Python 3.10+	Linux, Windows, or macOS
Nvidia GPU (Apple M chips may work, but Nvidia GPU is highly recommended)	
Hugging Face Diffusers	

Please turn to *Chapter 2* for detailed steps to set up the environment.

If you are using the digital version of this book, we advise you to type the code yourself or access the code from the book's GitHub repository (a link is available in the next section). Doing so will help you avoid any potential errors related to the copying and pasting of code.

Download the example code files

You can download the example code files for this book from GitHub at `https://github.com/PacktPublishing/Using-Stable-Diffusion-with-Python`. If there's an update to the code, it will be updated in the GitHub repository.

We also have other code bundles from our rich catalog of books and videos available at `https://github.com/PacktPublishing/`. Check them out!

Conventions used

There are a number of text conventions used throughout this book.

`Code in text`: Indicates code words in text, database table names, folder names, filenames, file extensions, pathnames, dummy URLs, user input, and Twitter handles. Here is an example: "Here, let's use the `controlnet-openpose-sdxl-1.0` open pose ControlNet for SDXL."

A block of code is set as follows:

```
import torch
from diffusers import StableDiffusionPipeline
# load model
text2img_pipe = StableDiffusionPipeline.from_pretrained(
    "stablediffusionapi/deliberate-v2",
    torch_dtype = torch.float16
).to("cuda:0")
```

Any command-line input or output is written as follows:

```
$ pip install pandas
```

Bold: Indicates a new term, an important word, or words that you see onscreen. For instance, words in menus or dialog boxes appear in **bold**. Here is an example: "After clicking the **Run** button, the progress bar will appear in the position of the output textbox."

> **Tips or important notes**
> Appear like this.

Get in touch

Feedback from our readers is always welcome.

General feedback: If you have questions about any aspect of this book, email us at `customercare@packtpub.com` and mention the book title in the subject of your message.

Errata: Although we have taken every care to ensure the accuracy of our content, mistakes do happen. If you have found a mistake in this book, we would be grateful if you would report this to us. Please visit www.packtpub.com/support/errata and fill in the form.

Piracy: If you come across any illegal copies of our works in any form on the internet, we would be grateful if you would provide us with the location address or website name. Please contact us at copyright@packt.com with a link to the material.

If you are interested in becoming an author: If there is a topic that you have expertise in and you are interested in either writing or contributing to a book, please visit authors.packtpub.com.

Share Your Thoughts

Once you've read *Using Stable Diffusion with Python*, we'd love to hear your thoughts! Scan the QR code below to go straight to the Amazon review page for this book and share your feedback.

https://packt.link/r/1-835-08637-3

Your review is important to us and the tech community and will help us make sure we're delivering excellent quality content.

Download a free PDF copy of this book

Thanks for purchasing this book!

Do you like to read on the go but are unable to carry your print books everywhere?

Is your eBook purchase not compatible with the device of your choice?

Don't worry, now with every Packt book you get a DRM-free PDF version of that book at no cost.

Read anywhere, any place, on any device. Search, copy, and paste code from your favorite technical books directly into your application.

The perks don't stop there, you can get exclusive access to discounts, newsletters, and great free content in your inbox daily

Follow these simple steps to get the benefits:

1. Scan the QR code or visit the link below

https://packt.link/free-ebook/9781835086377

2. Submit your proof of purchase
3. That's it! We'll send your free PDF and other benefits to your email directly

Part 1 – A Whirlwind of Stable Diffusion

Welcome to the fascinating world of Stable Diffusion, a rapidly evolving field that has revolutionized the way we approach image generation and manipulation. In the first part of our journey, we'll embark on a comprehensive exploration of the fundamentals, laying the groundwork for a deep understanding of this powerful technology.

Over the next six chapters, we'll delve into the core concepts, principles, and applications of Stable Diffusion, providing a solid foundation for further experimentation and innovation. We'll begin by introducing the basics of Stable Diffusion, followed by a hands-on guide to setting up your environment for success. You'll then learn how to generate stunning images using Stable Diffusion, before diving deeper into the theoretical underpinnings of diffusion models and the intricacies of how Stable Diffusion works its magic.

By the end of this part, you'll possess a broad understanding of Stable Diffusion, from its underlying mechanics to practical applications, empowering you to harness its potential and create remarkable visual content. So, let's dive in and discover the wonders of Stable Diffusion!

This part contains the following chapters:

- *Chapter 1, Introducing Stable Diffusion*
- *Chapter 2, Setup Environment for Stable Diffusion*
- *Chapter 3, Generate Images Using Stable Diffusion*
- *Chapter 4, Understand the Theory behind the Diffusion Models*
- *Chapter 5, Understanding How Stable Diffusion Works*
- *Chapter 6, Using the Stable Diffusion Model*

1

Introducing Stable Diffusion

Stable Diffusion is a deep learning model that utilizes diffusion processes to generate high-quality artwork from guided instructions and images.

In this chapter, we will introduce you to AI image generation technology, namely Stable Diffusion, and see how it evolved into what it is now.

Unlike other deep learning image generation models, such as OpenAI's DALL-E 2, Stable Diffusion works by starting with a random-noise latent tensor and then gradually adding detailed information to it. The amount of detail that is added is determined by a diffusion process, governed by a mathematical equation (we will delve into the details in *Chapter 5*). In the final stage, the model decodes the latent tensor into the pixel image.

Since its creation in 2022, Stable Diffusion has been used widely to generate impressive images. For example, it can generate images of people, animals, objects, and scenes that are indistinguishable from real photographs. Images are generated using specific instructions, such as *A cat running on the moon's surface* or *a photograph of an astronaut riding a horse*.

Here is a sample of a prompt to use with Stable Diffusion to generate an image using the given description:

```
"a photograph of an astronaut riding a horse".
```

Stable Diffusion will generate an image like the following:

Figure 1.1: A photograph of an astronaut riding a horse, generated by Stable Diffusion

This image didn't exist before I hit the *Enter* button. It was created collaboratively by me and Stable Diffusion. Stable Diffusion not only understands the descriptions we give it, but also adds more detail to the image.

Apart from text-to-image generation, Stable Diffusion also facilitates editing photos using natural language. To illustrate, consider the preceding image again. We can replace the space background with a blue sky and mountains using an automatically generated mask and prompts.

The `background` prompt can be used to generate the background mask, and the `blue sky and mountains` prompt is used to guide Stable Diffusion to transform the initial image into the following:

Figure 1.2: Replace the background with a blue sky and mountains

No mouse-clicking or dragging is required, and there's no need for additional paid software such as Photoshop. You can achieve this using pure Python together with Stable Diffusion. Stable Diffusion can perform many other tasks using only Python code, which will be covered later in this book.

Stable Diffusion is a powerful tool that has the potential to revolutionize the way we create and interact with images. It can be used to create realistic images for movies, video games, and other applications. It can also be used to generate personalized images for marketing, advertising, and decoration.

Here are some of the key features of Stable Diffusion:

- It can generate high-quality images from text descriptions
- It is based on a diffusion process, which is a more stable and reliable way to generate images than other methods
- Many massive pre-trained publicly accessible models are available (10,000+), and keep on growing
- New research and applications are building on Stable Diffusion
- It is open source and can be used by anyone

Before we proceed, let me provide a brief introduction to the evolution of the Diffusion model in recent years.

Evolution of the Diffusion model

Diffusion hasn't always been available, just as Rome was not built in a day. To have a high-level bird's view of this technology, in this section, we will discuss the overall evolution of the Diffusion model in recent years.

Before Transformer and Attention

Not too long ago, **Convolutional Neural Networks** (**CNNs**) and **Residual Neural Networks** (**ResNets**) dominated the field of computer vision in machine learning.

CNNs and ResNets have proven to be highly effective in tasks such as guided object detection and face recognition. These models have been widely adopted across various industries, including self-driving cars and AI-driven agriculture.

However, there is a significant drawback to CNNs and ResNets: they can only recognize objects that are part of their training set. To detect a completely new object, a new category label must be added to the training dataset, followed by retraining or fine-tuning the pre-trained models.

This limitation stems from the models themselves, as well as the constraints imposed by hardware and the availability of training data at that time.

Transformer transforms machine learning

The Transformer model, developed by Google, has revolutionized the field of computer vision, starting with its impact on **Natural Language Processing (NLP)**.

Unlike traditional approaches that rely on predefined labels to calculate loss and update neural network weights through backpropagation, the Transformer model, along with the Attention mechanism, introduced a pioneering concept. They utilize the training data itself for both training and labeling purposes.

Let's consider the following sentence as an example:

"Stable Diffusion can generate images using text"

Let's say we input the sequence of words into the neural network, excluding the last word *text*:

"Stable Diffusion can generate images using"

Using this prompt, the model can predict the next word based on its current weights. Let's say it predicts *apple*. The encoded embedding of the word *apple* is significantly different from *text* in terms of vector space, much like two numbers with a large gap between them. This gap value can be used as the loss value, which is then backpropagated to update the weights.

By repeating this process millions or even billions of times during training and updating, the model's weights gradually learn to produce the next reasonable words in a sentence.

Machine learning models can now learn a wide range of tasks with a properly designed loss function.

CLIP from OpenAI makes a big difference

Researchers and engineers quickly recognized the potential of the Transformer model, as mentioned in the concluding remarks of the well-known machine learning paper titled *Attention Is All You Need* [2]. The author states the following:

We are excited about the future of Attention-based models and plan to apply them to other tasks. We plan to extend the Transformer to problems involving input and output modalities other than text and to investigate local, restricted Attention mechanisms to efficiently handle large inputs and outputs such as images, audio, and video.

If you have read the paper and grasped the remarkable capabilities of Transformer- and Attention-based models, you might also be inspired to reimagine your own work and harness this extraordinary power.

Researchers from OpenAI grasped this power and created a model called CLIP [1] that uses the Attention mechanism and Transformer model architecture to train an image classification model. The model has the ability to classify a wide range of images with no need for labeled data. It is the first large-scale image classification model trained on 400 million image-text pairs extracted from the internet.

Although there were similar efforts prior to OpenAI's CLIP model, the results were not deemed satisfactory according to the authors of the CLIP paper [1]:

A crucial difference between these weakly supervised models and recent explorations of learning image representations directly from natural language is scale.

Indeed, scale plays a pivotal role in unlocking the remarkable superpower of universal image recognition. While other models utilized 200,000 images, the CLIP team trained their model using a staggering 400,000,000 images combined with text data from the public internet.

The results are astonishing. CLIP enables image recognition and segmentation without the limitations of predefined labels. It can detect objects that previous models struggled with. CLIP has brought about a significant change through its large-scale model. Given the immense weight of CLIP, researchers have pondered whether it could also be employed for image generation from text.

Generate images

Using only CLIP, we still cannot generate a realistic image based on a text description. For instance, if we ask CLIP to draw an apple, the model merges various types of apples, different shapes, colors, backgrounds, and so on. CLIP might generate an apple that is half green and half red, which might not be what we intended.

You may be familiar with **Generative Adversarial Networks** (**GANs**), which are capable of generating highly photorealistic images. However, text prompts cannot be utilized in the generation process. GANs have become a sophisticated solution for image processing tasks such as face restoration and image upscaling. Nevertheless, a new innovative approach was needed to leverage models for image generation based on guided descriptions or prompts.

In June 2020, a paper titled *Denoising Diffusion Probabilistic Models* [3] by Jonathan Ho et al. introduced a diffusion-based probabilistic model for image generation. The term **diffusion** is borrowed from thermodynamics. The original meaning of diffusion is the movement of particles from a region of high concentration to a region of low concentration. This idea of diffusion inspired machine learning researchers to apply it to denoising and sampling processes. In other words, we can start with a noisy image and gradually refine it by removing noise. The denoising process gradually transforms an image with high levels of noise into a clearer version of the original image. Therefore, this generative model is referred to as a **denoising diffusion probabilistic model**.

The idea behind this approach is ingenious. For any given image, a limited number of normally distributed noise images are added to the original image, effectively transforming it into a fully noisy image. What if we train a model that can reverse this diffusion process, guided by the CLIP model? Surprisingly, this approach works [4].

DALL-E 2 and Stable Diffusion

In April 2022, OpenAI released DALL-E 2, accompanied by its paper titled *Hierarchical Text-Conditional Image Generation with CLIP Latents* [4]. DALL-E 2 garnered significant attention worldwide. It generated a massive collection of astonishing images that spread across social networks and mainstream media. People were not only amazed by the quality of the generated images but also by its ability to create images that had never existed before. DALL-E 2 was effectively producing works of art.

Perhaps coincidentally, in April 2022, a paper titled *High-Resolution Image Synthesis with Latent Diffusion Models* [5] was published by CompVis, introducing another diffusion-based model for text-guided image generation. Building upon CompVis's work, researchers and engineers from CompVis, Stability AI, and LAION collaborated to release an open source counterpart of DALL-E 2 called Stable Diffusion in August 2022.

Why Stable Diffusion

While DALL-E 2 and other commercial image generation models such as Midjourney can produce remarkable images without requiring complex environment setups or hardware preparation, these models are closed-source. Consequently, users have limited control over the generation process, cannot use their own customized models, and are unable to add custom functions to the platform.

On the other hand, Stable Diffusion is an open source model released under the CreativeML Open RAIL-M license. Users not only have the freedom to utilize the model but can also read the source code, add features, and benefit from the countless custom models shared by the community.

Which Stable Diffusion to use

When we say Stable Diffusion, which Stable Diffusion are we really referring to? Here's a list of the different Stable Diffusion tools and the differences between them:

- **Stable Diffusion GitHub repo** (`https://github.com/CompVis/stable-diffusion`): This is the original implementation of Stable Diffusion from CompVis, contributed to by many great engineers and researchers. It is a PyTorch implementation that can be used to train and generate images, text, and other creative content. The library is now less active at the time of writing in 2023. Its README page also recommends users use Diffusers from Hugging Face to use and train Diffusion models.

- **Diffusers from Hugging Face**: Diffusers is a library for training and using diffusion models developed by Hugging Face. It is the go-to library for state-of-the-art, pre-trained diffusion models for generating images, audio, and even the 3D structures of molecules. The library is well maintained and being actively developed at the time of writing. New code is added to its GitHub repository almost every day.

- **Stable Diffusion WebUI from AUTOMATIC1111**: This might be the most popular web-based application currently that allows users to generate images and text using Stable Diffusion. It provides a GUI interface that makes it easy to experiment with different settings and parameters.

- **InvokeAI**: InvokeAI was originally developed as a fork of the Stable Diffusion project, but it has since evolved into its own unique platform. InvokeAI offers a number of features that make it a powerful tool for creatives.

- **ComfyUI**: ComfyUI is a node-based UI that utilizes Stable Diffusion. It allows users to construct tailored workflows, including image post-processing and conversions. It is a potent and adaptable graphical user interface for Stable Diffusion, characterized by its node-based design.

In this book, when I use Stable Diffusion, I am referring to the Stable Diffusion model, not the GUI tools just listed. The focus of this book will be on using Stable Diffusion with plain Python. Our example code will use Diffusers' pipelines and will leverage the code from Stable Diffusion WebUI and open source code from academic papers, et cetera.

Why this book

While the Stable Diffusion GUI tool can generate fantastic images driven by the Diffusion model, its usability is limited. The presence of dozens of knobs (more sliders and buttons are being added) and specific terms sometimes makes generating high-quality images a guessing game. On the other hand, the open source Diffusers package from Hugging Face gives users full control over Stable Diffusion using Python. However, it lacks many key features such as loading custom LoRA and textual inversion, utilizing community-shared models/checkpoints, scheduling, and weighted prompts, unlimited prompt tokens, and high-resolution image fixing and upscaling (The Diffusers package does keep improving over time, however).

This book aims to help you understand all the complex terms and knobs from the internal view of the Diffusion model. The book will also assist you in overcoming the limitations of Diffusers and implementing the missing functions and advanced features to create a fully customized Stable Diffusion application.

Considering the rapid pace of AI technology evolution, this book also aims to enable you to quickly adapt to the upcoming changes.

By the end of this book, you will not only be able to use Python to generate and edit images but also leverage the solutions provided in the book to build Stable Diffusion applications for your business and users.

Let's start the journey.

References

1. *Learning Transferable Visual Models From Natural Language Supervision*: https://arxiv.org/abs/2103.00020

2. *Attention Is All You Need*: https://arxiv.org/abs/1706.03762

3. *Denoising Diffusion Probabilistic Models*: https://arxiv.org/abs/2006.11239

4. *Hierarchical Text-Conditional Image Generation with CLIP Latents*: https://arxiv.org/abs/2204.06125v1

5. *High-Resolution Image Synthesis with Latent Diffusion Models*: https://arxiv.org/abs/2112.10752

6. DALL-E 2: https://openai.com/dall-e-2

2

Setting Up the Environment for Stable Diffusion

Welcome to *Chapter 2*. In this chapter, we will be focusing on setting up the environment to run Stable Diffusion. We will cover all the necessary steps and aspects to ensure a seamless experience while working with Stable Diffusion models. Our primary goal is to help you understand the importance of each component and how they contribute to the overall process.

The contents of this chapter are as follows:

- Introduction to the hardware requirements to run Stable Diffusion
- Detailed steps to install the required software dependencies: CUDA from NVIDIA, Python, a Python virtual environment (optional but recommended), and PyTorch
- Alternative options for users without a GPU, such as Google Colab and Apple MacBook with silicon CPU (M series)
- Troubleshooting common issues during the setup process
- Tips and best practices for maintaining a stable environment

We will begin by providing an overview of Stable Diffusion, its significance, and its applications in various fields. This will help you gain a better understanding of the core concept and its importance.

Next, we will dive into the step-by-step installation process for each dependency, including CUDA, Python, and PyTorch. We will also discuss the benefits of using a Python virtual environment and guide you through setting one up.

For those who do not have access to a machine with a GPU, we will explore alternative options such as Google Colab. We will provide a comprehensive guide to using these services and discuss the trade-offs associated with them.

Finally, we will address common issues that may arise during the setup process and provide troubleshooting tips. Additionally, we will share best practices for maintaining a stable environment to ensure a smooth experience while working with Stable Diffusion models.

By the end of this chapter, you will have a solid foundation for setting up and maintaining an environment tailored for Stable Diffusion, allowing you to focus on building and experimenting with your models efficiently.

Hardware requirements to run Stable Diffusion

This section will discuss the hardware requirements of running a Stable Diffusion model. This book will cover **Stable Diffusion v1.5** and the **Stable Diffusion XL** (**SDXL**) version. These two are also the most used models at the time of writing this book.

Stable Diffusion v1.5, released in October 2022, is considered a general-purpose model, and can be used interchangeably with v1.4. On the other hand, SDXL, which was released in July 2023, is known for its ability to handle higher resolutions more effectively compared to Stable Diffusion v1.5. It can generate images with larger dimensions without compromising on quality.

Essentially, Stable Diffusion is a set of models that includes the following:

- **Tokenizer**: This tokenizes a text prompt into a sequence of tokens.
- **Text Encoder**: The Stable Diffusion text encoder is a special Transformer language model – specifically, the text encoder of a CLIP model. In SDXL, a larger-size OpenCLIP [6] text encoder is also used to encode the tokens into text embeddings.
- **Variational Autoencoder** (**VAE**): This encodes images into a latent space and decodes them back into images.
- **UNet**: This is where the denoising process happens. The UNet structure is employed to comprehend the steps involved in the noising/denoising cycle. It accepts certain elements such as noise, time step data, and a conditioning signal (for instance, a representation of a text description), and forecasts noise residuals that can be utilized in the denoising process.

The components of Stable Diffusion provide neural network weight data, except for the tokenizer. While the CPU can handle the training and inference in theory, a physical machine with a GPU or parallel computing device can provide the best experience to learn and run Stable Diffusion models.

GPU

In theory, Stable Diffusion models can run on both GPU and CPU. In reality, PyTorch-based models work best on an NVIDIA GPU with CUDA.

Stable Diffusion requires a GPU with at least 4 GB VRAM. From my own experience, a GPU with 4 GB VRAM can only enable you to generate 512x512 images but it may take a long time to generate them. A GPU with at least 8 GB VRAM grants a relatively pleasant learning and usage experience. The larger the VRAM size, the better.

The code of this book is tested on NVIDIA RTX 3070Ti with 8 GB VRAM and RTX 3090 with 24 GB VRAM.

System memory

There will be a lot of data transferred between GPU and CPU, and some Stable Diffusion models will easily take up to 6 GB RAM. Please prepare at least 16 GB of system RAM; 32 GB RAM will be good – the more, the better, especially for multiple models.

Storage

Do prepare a large drive. By default, the Hugging Face package will download model data to a cache folder located in the system drive. If you only have 256 GB or 512 GB storage, you will find it quickly running out. Preparing a 1 TB NVME SSD is recommended, although 2 TB or more will be even better.

Software requirements

Now we have the hardware prepared, Stable Diffusion requires additional software to support its execution and provide better control using Python. This section will provide you with the steps to prepare the software environment.

CUDA installation

If you are using Microsoft Windows, please install Microsoft **Visual Studio** (**VS**) [5] first. VS will install all other dependent packages and binary files for CUDA. You can simply choose the latest Community version of VS for free.

Now, go to the NVIDIA CUDA download page [1] to get the CUDA installation file. The following screenshot shows an example of selecting CUDA for Windows 11:

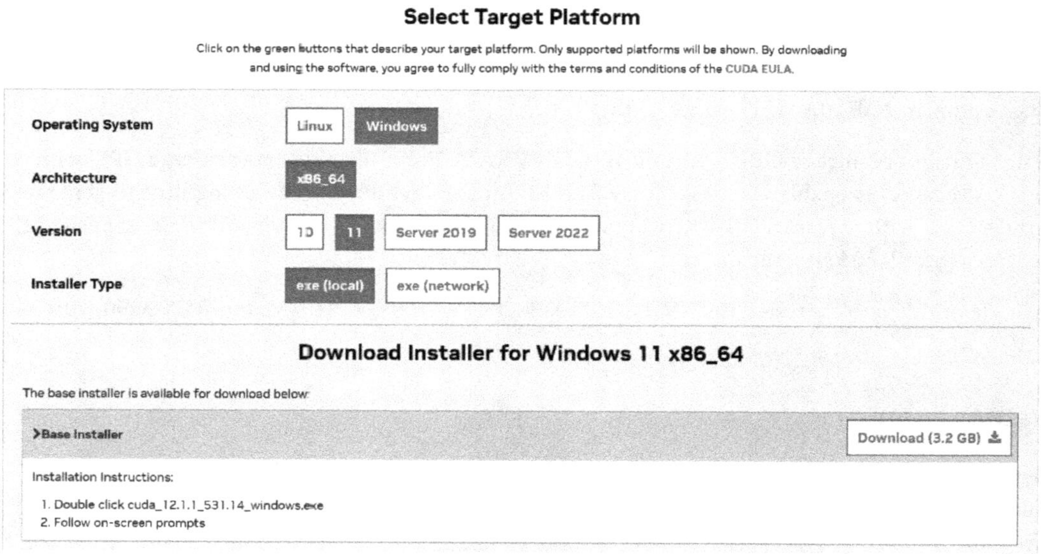

Select Target Platform

Click on the green buttons that describe your target platform. Only supported platforms will be shown. By downloading and using the software, you agree to fully comply with the terms and conditions of the CUDA EULA.

Operating System	Linux	Windows		
Architecture	x86_64			
Version	10	11	Server 2019	Server 2022
Installer Type	exe (local)	exe (network)		

Download Installer for Windows 11 x86_64

The base installer is available for download below:

>Base Installer	Download (3.2 GB) ⬇

Installation Instructions:

1. Double click cuda_12.1.1_531.14_windows.exe
2. Follow on-screen prompts

Figure 2.1: Selecting the CUDA installation download file for Windows

Download the CUDA installation file, then double-click this file to install CUDA like any other Windows application.

If you are using a Linux operating system, installing CUDA for Linux is slightly different. You can execute the Bash script provided by NVIDIA to automate the installation. Here are the detailed steps:

1. It is better to uninstall all NVIDIA drivers first to ensure minimum errors, so if you have NVIDIA's driver already installed, use the following command to uninstall it:

    ```
    sudo apt-get purge nvidia*
    sudo apt-get autoremove
    ```

    Then, reboot your system:

    ```
    sudo reboot
    ```

2. Install GCC. **GNU Compiler Collection (GCC)** is a set of compilers for various programming languages such as C, C++, Objective-C, Fortran, Ada, and others. It is an open source project developed by the GNU Project and is widely used for compiling and building software on Unix-like operating systems, including Linux. Without GCC being installed, we will get errors during the CUDA installation. Install it with the following command:

    ```
    sudo apt install gcc
    ```

3. Select the right CUDA version for your system on the CUDA download page [2]. The following screenshot shows an example of selecting CUDA for Ubuntu 22.04:

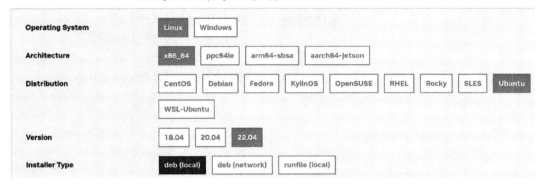

Figure 2.2: Selecting the CUDA installation download file for Linux

After your selection, the page will show you the command scripts that handle the entire installation process. Here is one example:

```
wget https://developer.download.nvidia.com/compute/cuda/repos/
ubuntu2204/x86_64/cuda-ubuntu2204.pin
sudo mv cuda-ubuntu2204.pin /etc/apt/preferences.d/cuda-repository-
pin-600
wget https://developer.download.nvidia.com/compute/cuda/12.1.1/local_
installers/cuda-repo-ubuntu2204-12-1-local_12.1.1-530.30.02-1_amd64.
deb
sudo dpkg -i cuda-repo-ubuntu2204-12-1-local_12.1.1-530.30.02-1_amd64.
deb
sudo cp /var/cuda-repo-ubuntu2204-12-1-local/cuda-*-keyring.gpg /usr/
share/keyrings/
sudo apt-get update
sudo apt-get -y install cuda
```

> **Note**
> The script may have been updated by the time you read this book. To avoid errors and potential installation failures, I would suggest opening the page and using the script that reflects your selection.

Installing Python for Windows, Linux, and macOS

We will first install Python for Windows.

Installing Python for Windows

You can visit https://www.python.org/ and download Python 3.9 or Python 3.10 to install it.

After years of manually downloading and clicking through the installation process, I found that using a package manager is quite useful to automate the installation. With a package manager, you write a script once, save it, and then the next time you need to install the software, all you have to do is run the same script in a terminal window. One of the best package managers for Windows is Chocolatey (https://chocolatey.org/).

Once you have Chocolatey installed, use the following command to install Python 3.10.6:

```
choco install python --version=3.10.6
```

Create a Python virtual environment:

```
pip install --upgrade --user pip
pip install virtualenv
python -m virtualenv venv_win_p310
venv_win_p310\Scripts\activate
python -m ensurepip
python -m pip install --upgrade pip
```

We will move on to the steps to install Python for Linux.

Installing Python for Linux

Let's now install Python for Linux (Ubuntu). Follow these steps:

1. Install the required packages:

    ```
    sudo apt-get install software-properties-common
    sudo add-apt-repository ppa:deadsnakes/ppa
    sudo apt-get update
    sudo apt-get install python3.10
    sudo apt-get install python3.10-dev
    sudo apt-get install python3.10-distutils
    ```

2. Install pip:

    ```
    curl https://bootstrap.pypa.io/get-pip.py -o get-pip.py
    python3.10 get-pip.py
    ```

3. Create a Python virtual environment:

    ```
    python3.10 -m pip install --user virtualenv
    python3.10 -m virtualenv venv_ubuntu_p310
    . venv_ubuntu_p310/bin/activate
    ```

Installing Python for macOS

If you are using a Mac with the silicon chip inside (with Apple Mx CPU), there is a high chance you have Python installed already. You can test whether you have Python installed on your Mac with the following command:

```
python3 --version
```

If your machine doesn't have a Python interpreter yet, you can install it with one simple command using Homebrew [7] like this:

```
brew install python
```

Keep in mind that Python versions are regularly updated, usually on an annual basis. You can change the version number to install a specific Python version. For example, you can change `python3.10` to `python3.11`.

Installing PyTorch

The Hugging Face Diffusers package relies on the PyTorch package, so we will need to have the PyTorch package installed. Go to the PyTorch **Get Started** page (https://pytorch.org/get-started/locally/) and select the appropriate PyTorch version for your system. The following is a screenshot of PyTorch for Windows:

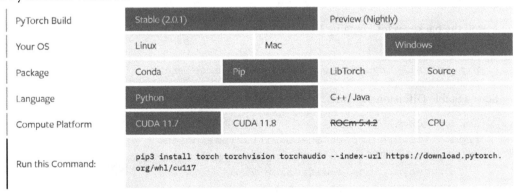

PyTorch Build	Stable (2.0.1)		Preview (Nightly)	
Your OS	Linux	Mac	Windows	
Package	Conda	Pip	LibTorch	Source
Language	Python		C++ / Java	
Compute Platform	CUDA 11.7	CUDA 11.8	ROCm 5.4.2	CPU
Run this Command:	pip3 install torch torchvision torchaudio --index-url https://download.pytorch.org/whl/cu117			

Figure 2.3: Installing PyTorch

Next, use the dynamically generated command to install PyTorch:

```
pip3 install torch torchvision torchaudio --index-url https://
download.pytorch.org/whl/cu117
```

In addition to CUDA 11.7, there is also CUDA 11.8. The choice of version will depend on the CUDA version installed on your machine.

You can use the following command to find out your CUDA version:

```
nvcc --version
```

You can also use this command:

```
nvidia-smi
```

Your machine's CUDA version may be higher than the listed versions of 11.7 or 11.8, such as 12.1. Often, a specific version is required by a certain model or package. For Stable Diffusion, just install the newest version.

If you are using a Mac, select the **Mac** option to install PyTorch for macOS.

If you are using a Python virtual environment, make sure to install PyTorch within the activated virtual environment. Otherwise, you may encounter issues where PyTorch is not installed correctly if you accidentally install it outside the virtual environment and then run your Python code within the virtual environment.

Running a Stable Diffusion pipeline

Now that you have installed all the dependencies, it is time to run a Stable Diffusion pipeline to test whether the environment is correctly set up. You can use any Python editing tool, such as VS Code or Jupyter Notebook, to edit and execute Python code. Follow these steps:

1. Install the packages for Hugging Face Diffusers:

    ```
    pip install diffusers
    pip install transformers scipy ftfy accelerate
    ```

2. Start a Stable Diffusion pipeline:

    ```
    import torch
    from diffusers import StableDiffusionPipeline
    pipe = StableDiffusionPipeline.from_pretrained(
        "runwayml/stable-diffusion-v1-5",
        torch_dtype=torch.float16)
    pipe.to("cuda") # mps for mac
    ```

    If you are using a Mac, change cuda to mps. Even though macOS is supported and can generate images using the Diffusers package, its performance is relatively slow. As a comparison, an NVIDIA RTX 3090 can achieve about 20 iterations per second to generate one 512x512 image using Stable Diffusion V1.5, whereas an M3 Max CPU can only reach around 5 iterations per second with the default settings.

3. Generate an image:

```
prompt = "a photo of an astronaut riding a horse on mars,blazing
fast, wind and sand moving back"
image = pipe(
    prompt, num_inference_steps=30
).images[0]
image
```

If you see an image of an astronaut riding a horse, you have all the environments set up correctly in your physical machine.

Using Google Colaboratory

Google Colaboratory (or **Google Colab**) is an online computing service provided by Google. In essence, Google Colab is an online Jupyter notebook with GPU/CUDA capability.

Its free notebook can provide CUDA computing with 15 GB VRAM equivalent to an NVIDIA RTX 3050 or RTX 3060. The performance is decent if you don't have a discrete GPU at hand.

Let's look at the advantages and disadvantages of using Google Colab:

- **Advantages**:

  - No need to manually install CUDA and Python

  - Everything is online; you can save a link and reopen it anywhere

  - The installation of pip and downloading of resources are fast

- **Disadvantages**:

  - There is a disk limitation for each notebook

  - You don't have full control of the backend server; terminal access requires a Colab Pro subscription

  - The performance is not guaranteed so you may experience slow GPU inference during peak time and could be disconnected for long-time computing

  - The Colab notebook compute environment will be reset every time you restart the notebook; in other words, you will need to reinstall all the packages and download the model files every time you start the notebook

Using Google Colab to run a Stable Diffusion pipeline

Here are the detailed steps to start using Google Colab:

1. Create a new instance from `https://colab.research.google.com/`.

2. Click **Runtime | Change runtime type** and select **T4 GPU**, as shown in *Figure 2.4*:

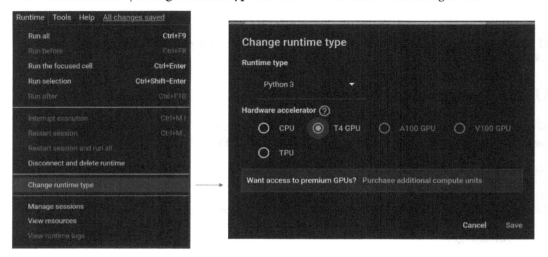

Figure 2.4: Selecting GPU in the Google Colab notebook

3. Create a new cell and use the following command to check whether the GPU and CUDA are working:

    ```
    !nvidia-smi
    ```

4. Install the packages for Hugging Face Diffusers:

    ```
    !pip install diffusers
    !pip install transformers scipy ftfy accelerate ipywidgets
    ```

5. Start a Stable Diffusion pipeline:

    ```
    import torch
    from diffusers import StableDiffusionPipeline
    pipe = StableDiffusionPipeline.from_pretrained(
        "runwayml/stable-diffusion-v1-5",
        torch_dtype=torch.float16)
    pipe.to("cuda")
    ```

6. Generate an image:

```
prompt = "a photo of an astronaut riding a horse on mars,blazing
fast, wind and sand moving back"
image = pipe(
    prompt, num_inference_steps=30
).images[0]
image
```

In a few seconds, you should be able to see the result as shown in *Figure 2.5*:

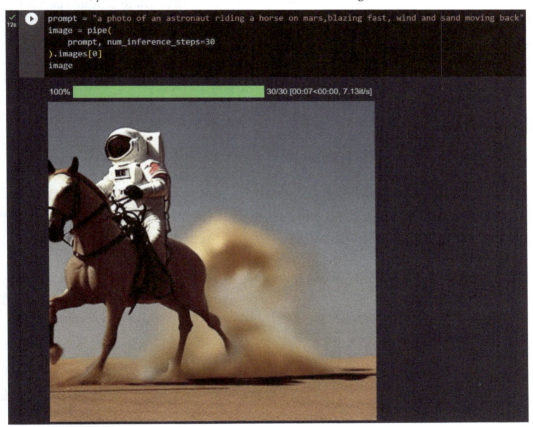

Figure 2.5: Running a Stable Diffusion pipeline in Google Colab

If you see an image generated as in *Figure 2.5*, you have successfully set up the Diffusers package to run the Stable Diffusion model in Google Colab.

Summary

Some say that the most challenging part of starting to train a machine learning model is not the math or its internal logic. Often, the biggest hurdle is setting up a proper working environment to run the model. It's not uncommon to see engineers and professors spend an entire weekend trying to install CUDA on their lab machines. This can be due to missing dependencies, skipped steps, or version incompatibilities.

I dedicated an entire chapter to covering the installation process, hoping that these detailed steps would help you avoid common pitfalls. By following these steps, you'll be able to delve into the Stable Diffusion model and start image generation with minimum obstacles.

Furthermore, the software and packages you installed will also work for Transformer-based large language models.

In the next chapter, we will start using Stable Diffusion to generate images.

References

1. *CUDA Installation Guide for Microsoft Windows*: https://docs.nvidia.com/cuda/ cuda-installation-guide-microsoft-windows/index.html

2. *NVIDIA CUDA Downloads*: https://developer.nvidia.com/cuda-downloads

3. *Google Colab*: https://colab.research.google.com/

4. *Hugging Face Diffusers Installation*: https://huggingface.co/docs/diffusers/ installation

5. *Visual Studio Community Download*: https://visualstudio.microsoft.com/ vs/community/

6. *OpenCLIP GitHub repository*: https://github.com/mlfoundations/open_clip

7. *Homebrew*: https://brew.sh/

3

Generating Images Using Stable Diffusion

In this chapter, we will start using common Stable Diffusion functionalities by leveraging the Hugging Face Diffusers package (`https://github.com/huggingface/diffusers`) and open-source packages. As we mentioned in *Chapter 1, Introduction to Stable Diffusion*, Hugging Face Diffusers is currently the most widely used Python implementation of Stable Diffusion. As we explore image generation, we will walk through the common terminologies used.

Assume you have all the packages and dependencies installed; if you see an error message saying no GPU is found or CUDA is required, refer to *Chapter 2* to set up the environment to run Stable Diffusion.

With this chapter, I aim to familiarize you with Stable Diffusion by using the Diffusers package from Hugging Face. We will dig into the internals of Stable Diffusion in the next chapter.

In this chapter, we will cover the following topics:

- How to log in to Hugging Face with Hugging Face tokens
- Generating an image using Stable Diffusion
- Using a generation seed to reproduce an image
- Using the Stable Diffusion scheduler
- Swapping or changing a Stable Diffusion model
- Using a guidance scale

Let's start.

Logging in to Hugging Face

You may use the `login()` function in the `huggingface_hub` library like this:

```
from huggingface_hub import login
login()
```

In doing so, you are authenticating with the Hugging Face Hub. This allows you to download pre-trained diffusion models that are hosted on the Hub. Without logging in, you may not be able to download these models using the model ID, such as `runwayml/stable-diffusion-v1-5`.

When you run the preceding code, you are providing your Hugging Face token. You may wonder about the steps to *access* the token, but don't worry. The token input dialog will provide links and information to *access* the token.

After you have logged in, you can download pre-trained diffusion models by using the `from_pretrained()` function in the Diffusers package. For example, the following code will download the `stable-diffusion-v1-5` model from the Hugging Face Hub:

```
import torch
from diffusers import StableDiffusionPipeline

text2img_pipe = StableDiffusionPipeline.from_pretrained(
    "runwayml/stable-diffusion-v1-5",
    torch_dtype = torch.float16
).to("cuda:0")
```

> **Note**
>
> You may have noticed that I am using `to("cuda:0")` instead of `to("cuda")` because in the case of multiple-GPU scenarios, you can change the CUDA index to tell Diffusers to use a specified GPU. For instance, you can use `to("cuda:1")` to use the second CUDA-enabled GPU to generate Stable Diffusion images.

After downloading the model, it is time to generate an image using Stable Diffusion.

Generating an image

Now that we have the Stable Diffusion model loaded up to the GPU, let's generate an image. `text2img_pipe` holds the pipeline object; all we need to provide is a `prompt` string, using natural language to describe the image we want to generate, as shown in the following code:

```
# generate an image
prompt ="high resolution, a photograph of an astronaut riding a horse"
image = text2img_pipe(
```

```
    prompt = prompt
).images[0]
image
```

Feel free to change the prompt to anything else that comes to your mind when you are reading this, for example, `high resolution, a photograph of a cat running on the surface of Mars` or `4k, high quality image of a cat driving a plane`. It is amazing how Stable Diffusion can generate images according to a description in purely natural language.

If you run the preceding code without changing it, you may see an image like this showing up:

Figure 3.1: An image of an astronaut riding a horse

I said *you may see an image like this* because there is a 99.99% chance you will not see the same image; instead, you will see an image with a similar look and feel. To make the generation consistent, we will need another parameter, called `generator`.

Generation seed

In Stable Diffusion, a seed is a random number that is used to initialize the generation process. The seed is used to create a noise tensor, which is then used by the diffusion model to generate an image. The same seed together with the same prompt and settings will generally produce the same image.

The generation seed is needed for two reasons:

- **Reproducibility**: By using the same seed, you can consistently generate the same image with identical settings and prompts.

- **Exploration**: You can discover diverse image variations by altering the seed number. This often leads to the emergence of novel and intriguing images.

When a seed number is not provided, the Diffusers package automatically generates a random number for each image creation process. However, you have the option to specify your preferred seed number, as demonstrated in the following Python code:

```
my_seed = 1234
generator = torch.Generator("cuda:0").manual_seed(my_seed)
prompt ="high resolution, a photograph of an astronaut riding a horse"
image = text2img_pipe(
    prompt = prompt,
    generator = generator
).images[0]
display(image)
```

In the preceding code, we use `torch` to create a `torch.Generator` object with a manual seed provided. We specifically use this generator for image generation. By doing this, we can reproduce the same image repeatedly.

The generation seed is one method to control Stable Diffusion image generation. Next, let's explore the scheduler for further customization.

Sampling scheduler

After discussing the generation seed, let's now delve into another essential aspect of Stable Diffusion image generation: the sampling scheduler.

The original Diffusion models have demonstrated impressive results in generating images. However, one drawback is the slow reverse-denoising process, which typically requires 1,000 steps to transform a random noise data space into a coherent image (specifically, latent data space, a concept we will explore further in *Chapter 4*). This lengthy process can be burdensome.

To shorten the image generation process, several solutions have been brought out by researchers. The idea is simple: instead of denoising 1,000 steps, what if we could take a sample and only perform the key steps on that sample? And this idea works. Samplers or schedulers enable the Diffusion model to generate an image in a mere 20 steps!

In the Hugging Face Diffusers package, these helpful components are referred to as **schedulers**. However, you may also encounter the term **sampler** in other resources. You may take a look at the Diffusers *Schedulers* [2] page for the latest supported schedulers.

By default, the Diffusers package uses `PNDMScheduler`. We can find it by running this line of code:

```
# Check out the current scheduler
text2img_pipe.scheduler
```

The code will return an object like this:

```
PNDMScheduler {
  "_class_name": "PNDMScheduler",
  "_diffusers_version": "0.17.1",
  "beta_end": 0.012,
  "beta_schedule": "scaled_linear",
  "beta_start": 0.00085,
  "clip_sample": false,
  "num_train_timesteps": 1000,
  "prediction_type": "epsilon",
  "set_alpha_to_one": false,
  "skip_prk_steps": true,
  "steps_offset": 1,
  "trained_betas": null
}
```

At first glance, the PNDMScheduler object's fields might seem complex and unfamiliar. However, as you delve deeper into the internals of the Stable Diffusion model in *Chapters 4* and *5*, these fields will become more familiar and comprehensible. The learning journey ahead promises to unravel the intricacies of the Stable Diffusion model and shed light on the purpose and significance of each field within the PNDMScheduler object.

Many list schedulers can generate images in as few as 20 to 50 steps. Based on my experience, the Euler scheduler is one of the top choices. Let's apply the Euler scheduler to generate an image:

```
from diffusers import EulerDiscreteScheduler
text2img_pipe.scheduler = EulerDiscreteScheduler.from_config(
    text2img_pipe.scheduler.config)
generator = torch.Generator("cuda:0").manual_seed(1234)
prompt ="high resolution, a photograph of an astronaut riding a horse"
image = text2img_pipe(
    prompt = prompt,
    generator = generator
).images[0]
display(image)
```

You can customize the number of denoising steps by using the num_inference_steps parameter. A higher step count generally leads to better image quality. Here, we set the scheduling steps to 20 and compared the results of the default PNDMScheduler and EulerDiscreteScheduler:

```
# Euler scheduler with 20 steps
from diffusers import EulerDiscreteScheduler
text2img_pipe.scheduler = EulerDiscreteScheduler.from_config(
    text2img_pipe.scheduler.config)
```

```
generator = torch.Generator("cuda:0").manual_seed(1234)
prompt ="high resolution, a photograph of an astronaut riding a horse"
image = text2img_pipe(
    prompt = prompt,
    generator = generator,
    num_inference_steps = 20
).images[0]
display(image)
```

The following figure shows the difference between the two schedulers:

Figure 3.2: Left: Euler scheduler with 20 steps; right: PNDMScheduler with 20 steps

In this comparison, the Euler scheduler correctly generates an image with all four horse legs, while the PNDM scheduler provides more detail but misses one horse leg. These schedulers perform remarkably well, reducing the entire image generation process from 1,000 steps to just 20 steps, making it feasible to run Stable Diffusion on home computers.

Note that each scheduler has advantages and disadvantages. You may need to try out the schedulers to find out which one fits the best.

Next, let's explore the process of replacing the original Stable Diffusion model with a community-contributed, fine-tuned alternative.

Changing a model

At the time of writing this chapter, there are numerous models available, fine-tuned based on the V1.5 Stable Diffusion model, contributed by the thriving user community. If the model file is hosted on Hugging Face, you can easily switch to a different model by changing its identifier, as shown in the following code snippet:

```
# Change model to "stablediffusionapi/deliberate-v2"
from diffusers import StableDiffusionPipeline
text2img_pipe = StableDiffusionPipeline.from_pretrained(
    "stablediffusionapi/deliberate-v2",
    torch_dtype = torch.float16
).to("cuda:0")

prompt ="high resolution, a photograph of an astronaut riding a horse"
image = text2img_pipe(
    prompt = prompt
).images[0]
display(image)
```

Additionally, you can also use a ckpt/safetensors model downloaded from civitai.com (http://civitai.com). Here, we demonstrate loading the deliberate-v2 model using the following code:

```
from diffusers import StableDiffusionPipeline
text2img_pipe = StableDiffusionPipeline.from_single_file(
    "path/to/deliberate-v2.safetensors",
    torch_dtype = torch.float16
).to("cuda:0")

prompt ="high resolution, a photograph of an astronaut riding a horse"
image = text2img_pipe(
    prompt = prompt
).images[0]
display(image)
```

The primary difference when loading a model from a local file lies in the use of the from_single_file function instead of from_pretrained. A ckpt model file can be loaded up using the preceding code.

In *Chapter 6* of this book, we will focus exclusively on model loading, covering both Hugging Face and local storage methods. By experimenting with various models, you can discover improvements, unique artistic styles, or better compatibility for specific use cases.

We have touched on the generation seed, scheduler, and model usage. Another parameter that plays a key role is guidance_scale. Let's take a look at it next.

Guidance scale

Guidance scale or **Classifier-Free Guidance (CFG)** is a parameter that controls the adherence of the generated image to the text prompt. A higher guidance scale will force the image to be more aligned with the prompt, while a lower guidance scale will give more space for Stable Diffusion to decide what to put into the image.

Here is a sample of applying a different guidance scale while keeping other parameters the same:

```
import torch
generator = torch.Generator("cuda:0").manual_seed(123)

prompt = """high resolution, a photograph of an astronaut riding a
horse on mars"""

image_3_gs = text2img_pipe(
    prompt = prompt,
    num_inference_steps = 30,
    guidance_scale = 3,
    generator = generator
).images[0]

image_7_gs = text2img_pipe(
    prompt = prompt,
    num_inference_steps = 30,
    guidance_scale = 7,
    generator = generator
).images[0]

image_10_gs = text2img_pipe(
    prompt = prompt,
    num_inference_steps = 30,
    guidance_scale = 10,
    generator = generator
).images[0]

from diffusers.utils import make_image_grid
images = [image_3_gs,image_7_gs,image_10_gs]
make_image_grid(images,rows=1,cols=3)
```

Figure 3.3 provides a side-by-side comparison:

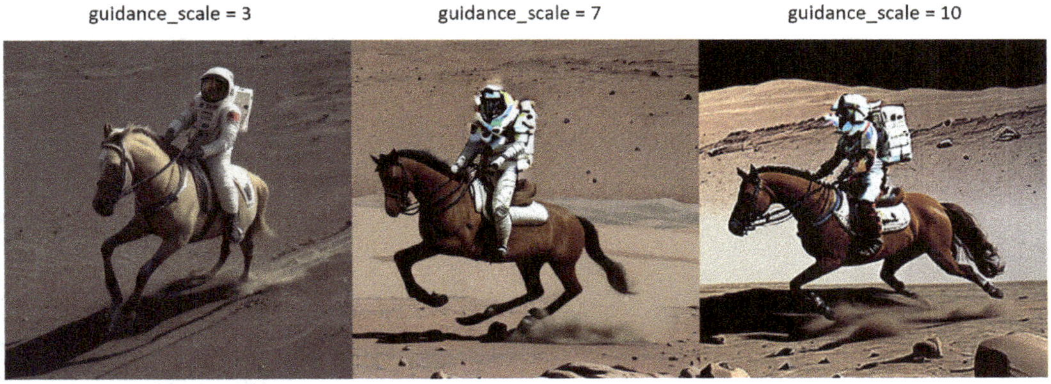

guidance_scale = 3   guidance_scale = 7   guidance_scale = 10

Figure 3.3: Left: guidance_scale = 3; middle: guidance_scale = 7; right: guidance_scale = 10

In practice, besides prompt adherence, we can notice that a high guidance scale setting has the following effects:

- Increases the color saturation
- Increases the contrast
- May lead to a blurred image if set too high

The `guidance_scale` parameter is typically set between 7 and 8.5. A value of 7.5 is a good default value.

Summary

In this chapter, we explored the essentials of using Stable Diffusion through the Hugging Face Diffusers package. We accomplished the following:

- Logged in to Hugging Face to enable automatic model downloads
- Generated images deterministically using the generator
- Utilized the scheduler for efficient image creation
- Adjusted the guidance scale for desired image qualities

With just a few lines of code, we successfully created images, demonstrating the remarkable capabilities of the Diffusers package. This chapter did not cover every feature and option; keep in mind that the package is continually evolving, with new functions and enhancements regularly added.

For those eager to unlock the full potential of the Diffusers package, I encourage you to explore its source code. Dive into the inner workings, uncover hidden gems, and build a Stable Diffusion pipeline from scratch. A rewarding journey awaits!

```
git clone https://github.com/huggingface/diffusers
```

In the next chapter, we will delve into the internals of the package and learn how to construct a custom Stable Diffusion pipeline tailored to your unique needs and preferences.

References

1. *High-Resolution Image Synthesis with Latent Diffusion Models*: https://arxiv.org/abs/2112.10752

2. Hugging Face Diffusers schedulers: https://huggingface.co/docs/diffusers/api/schedulers/overview

4

Understanding the Theory Behind Diffusion Models

This chapter will dive into the theory that powers **diffusion models** and see the internal workings of the system. How could a neural network model generate such realistic images? Curious minds would like to lift the cover and see the internal workings.

We are going to touch on the foundation of the diffusion model, aiming to figure out how it works internally and pave the foundation to implement a workable pipeline in the next chapter.

By comprehending the intricacies of diffusion models, we not only enhance our understanding of the advanced **Stable Diffusion** (also known as **latent diffusion models (LDMs)**) but also gain the ability to navigate the source code of the Diffusers package more effectively.

This knowledge will enable us to extend the package's features in line with emerging requirements.

Specifically, we will go through the following topics:

- Understanding the image-to-noise process
- A more efficient **forward diffusion process**
- The noise-to-image training process
- The noise-to-image sampling process
- Understanding Classifier Guidance denoising

By the end of this chapter, we will have taken a deep dive into the internal workings of the diffusion model initially brought out by Jonathan Ho et al. [4]. We will understand the foundational idea of the diffusion model and learn about the **forward diffusion process**. We will understand the reverse diffusion process for diffusion model training and sampling and learn to enable a text-guided diffusion model.

Let's get started.

Understanding the image-to-noise process

The idea of the diffusion model is inspired by the diffusion concept from thermodynamics. Take one image as a cup of water and add enough noise (ink) to the image (water) to finally turn the image (water) into a complete noise image (ink water).

As shown in *Figure 4.1*, image x_0 can be converted to a nearly Gaussian (normally distributed) noise image x_T.

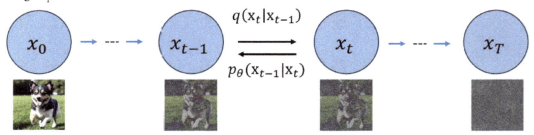

Figure 4.1: Forward diffusion and reverse denoising

We employ a predetermined forward diffusion process, denoted as q, which systematically introduces Gaussian noise to an image until it culminates in pure noise. The process is denoted by $q(x_t|x_{t-1})$. Note that the reverse process $p_\theta(x_{t-1}|x_t)$ is still unknown.

One step of the forward diffusion process can be denoted as follows:

$$q(x_t|x_{t-1}) := \mathcal{N}\left(x_t; \sqrt{1 - \beta_t x_{t-1}}, \beta_t I\right)$$

Let me explain this formula bit by bit from left to right:

- The notation $q(x_t|x_{t-1})$ is used to denote a conditional probability distribution. In this case, the distribution q represents the probability of observing the noisy image x_t given the previous image x_{t-1}.

- The define sign : = is used in the formula instead of the tilde symbol ($\sim$) because the diffusion forward process is a deterministic process. The tilde symbol ($\sim$) is typically used to represent a distribution. In this case, if we used the tilde symbol, the formula would be saying that the noisy image is a complete Gaussian distribution. However, this is not the case. The noisy image in t step is defined by a deterministic function of the previous image and added noise.

- Then why is $\mathcal{N}$ used here? The $\mathcal{N}$ symbol is used to represent a Gaussian distribution. However, in this case, the $\mathcal{N}$ symbol is being used to represent the functional form of the noisy image.

- On the right side, before the semicolon, x_t is the thing we want to have in normal distribution. After the semicolon, those are the parameters of the distribution. A semicolon is usually used to separate the output and parameters.

- β_t is the noise variance at step t. $\sqrt{1 - \beta_t} x_{t-1}$ is the mean of the new distribution.

- Why is the big I used in the formula? Because an RGB image can have multiple channels, and the identity matrix can apply the noise variance to different channels independently.

It is quite easy to add Gaussian noise to an image using Python:

```python
import numpy as np
import matplotlib.pyplot as plt
import ipyplot
from PIL import Image

# Load an image
img_path = r"dog.png"
image = plt.imread(img_path)

# Parameters
num_iterations = 16
beta = 0.1                  # noise_variance

images = []
steps = ["Step:"+str(i) for i in range(num_iterations)]

# Forward diffusion process
for i in range(num_iterations):
    mean = np.sqrt(1 - beta) * image
    image = np.random.normal(mean, beta, image.shape)

    # convert image to PIL image object
    pil_image = Image.fromarray((image * 255).astype('uint8'), 'RGB')

    # add to image list
    images.append(pil_image)

ipyplot.plot_images(images, labels=steps, img_width=120)
```

To execute the preceding code, you will also need to install the `ipyplot` package by `pip install ipyplot`. The code provided performs a simulation of a forward diffusion process on an image and then visualizes the progression of this process over a number of iterations. Here's a step-by-step explanation of what each part of the code is doing:

1. Importing libraries:

   - `ipyplot` is a library for plotting images in Jupyter notebooks in a more interactive way.

   - `PIL` (which stands for **Python Imaging Library**), specifically the `Image` module, is used for image manipulation.

2. Loading the image:

    - `img_path` is defined as the path to the image file `dog.png`.

    - `image` is loaded using `plt.imread(img_path)`.

3. Setting parameters:

    - `num_iterations` defines the number of times the diffusion process will be simulated.

    - `beta` is a parameter that simulates noise variance in the diffusion process.

4. Initializing lists:

    - `images` is initialized as an empty list, which will later hold the PIL image objects that result from each iteration of the diffusion process.

    - `steps` is a list of strings that will act as labels for the images when they are plotted, indicating the step number for each image.

5. Forward diffusion process:

    - A `for` loop runs for `num_iterations` times, each time performing a diffusion step. `mean` is computed by scaling the image with a factor of `sqrt(1 - beta)`.

    - A new image is generated by adding Gaussian noise to the mean, where the noise has a standard deviation of `beta`. This is done using `np.random.normal`.

    - The resulting image array values are scaled to the range 0-255 and converted to an 8-bit unsigned integer format, which is a common format for images.

    - `pil_image` is created by converting the image array to a PIL image object in RGB mode.

6. Plot the image using `ipyplot` in a grid as shown in *Figure 4.2*.

Figure 4.2: Add noise to the image

From the result, we can see that even though every image is from a normal distribution function, not every image is a complete Gaussian distribution, or more strictly speaking, an **isotropic Gaussian distribution**. The image will become a complete Gaussian distribution only when setting the step to infinite. But this is unnecessary. In the original DDPM paper [4], the step number is set to 1000, and later, in Stable Diffusion, the step number is reduced to between 20 to 50.

If the last image of *Figure 4.2* is an isotropic Gaussian distribution, its 2D distribution visualization will appear as a circle; it is characterized by having equal variances in all dimensions. In other words, the spread or width of the distribution is the same along all axes.

Let's plot an image pixel distribution after adding 16x times Gaussian noise:

```
sample_img = image  # take the last image from the diffusion process

plt.scatter(sample_img[:, 0], sample_img[:, 1], alpha=0.5)
plt.title("2D Isotropic Gaussian Distribution")
plt.xlabel("X")
plt.ylabel("Y")
plt.axis("equal")
plt.show()
```

The result is shown in *Figure 4.3*.

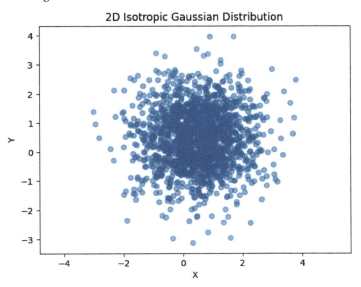

Figure 4.3: A nearly isotropic, normally distributed noise image

The figure shows how the code efficiently transforms an image into a nearly isotropic, normally distributed noise image in just 16 steps, as illustrated in the last image of *Figure 4.2*.

A more efficient forward diffusion process

If we use the chained process to calculate a noisy image at t step, it first requires calculating the noisy image from 1 to $t − 1$ steps, which is not efficient. We can leverage a trick called **reparameterization** [10] to transform the original chained process into a one-step process. Here is what the trick looks like.

If we have a Gaussian distribution z with μ as the mean and σ^2 variance:

$$z \sim \mathcal{N}(\mu, \sigma^2)$$

Then, we can rewrite the distribution as follows:

$$\epsilon \sim \mathcal{N}(0,1)$$

$$z = \mu + \sigma\epsilon$$

The benefit brought by this trick is that we can now calculate an image at any step with a one-step calculation, which will greatly boost the training performance:

$$x_t = \sqrt{1 − \beta_t}\, x_{t-1} + \sqrt{\beta_t}\, \epsilon_{t-1}$$

Now, say we define the following:

$$\alpha_t = 1 − \beta_t$$

We now have the following:

$$\overline{\alpha}_t = \prod_{i=1}^{t} \alpha_i$$

There is no magic here; define α_t and $\overline{\alpha}_t$ is only for convenience, so that we can calculate a noised image at step t and generate x_t from the source un-noised image x_0 using the following equation:

$$x_t = \sqrt{\overline{\alpha}_t}\, x_0 + \sqrt{1 − \overline{\alpha}_t}$$

What do α_t and $\overline{\alpha}_t$ look like? Here is a simplified sample (*Figure 4.4*).

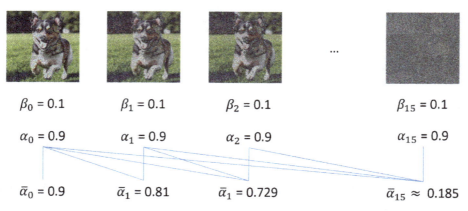

Figure 4.4: Implementation of reparameterization

In *Figure 4.4*, we have all the same α - 0.1 and β - 0.9. Now, whenever we need to generate a noised image x_t, we can quickly calculate $\overline{\alpha}_t$ from known numbers; the lines show what numbers are used to calculate $\overline{\alpha}_t$.

The following code can generate a noised image at any step:

```
import numpy as np
import matplotlib.pyplot as plt
from PIL import Image
from itertools import accumulate


def get_product_accumulate(numbers):
    product_list = list(accumulate(numbers, lambda x, y: x * y))
    return product_list


# Load an image
img_path = r"dog.png"
image = plt.imread(img_path)
image = image * 2 - 1                       # [0,1] to [-1,1]


# Parameters
num_iterations = 16
beta = 0.05                                 # noise_variance
betas = [beta]*num_iterations


alpha_list = [1 - beta for beta in betas]


alpha_bar_list = get_product_accumulate(alpha_list)


target_index = 5
x_target = (
    np.sqrt(alpha_bar_list[target_index]) * image
    + np.sqrt(1 - alpha_bar_list[target_index]) *
    np.random.normal(0,1,image.shape)
)


x_target = (x_target+1)/2


x_target = Image.fromarray((x_target * 255).astype('uint8'), 'RGB')
display(x_target)
```

This code is the implementation of the previously presented math formula. I present the code here to help build a correlated understanding between the math formula and the real implementation. If you are familiar with Python, you may find that this code makes the underlying subtleties easier to understand. The code can generate a noised image as shown in *Figure 4.5*.

Figure 4.5: Implementation of reparameterization

Now, let's think about how to recover an image by leveraging a neural network.

The noise-to-image training process

We have the solution to add noise to the image, which is known as forward diffusion, as shown in *Figure 4.6*. To recover an image from the noise, or **reverse diffusion**, as shown in *Figure 4.6*, we need to find a way to implement the reverse step $p_\theta(x_{t-1}|x_t)$. However, this step is intractable or uncomputable without additional help.

Consider that we have the ending Gaussian noise data, and all those noise step data in hand. What if we can train a neural network that can reverse the process? We can use the neural network to provide the mean and variance of a noise image and then remove the generated noise from the previous image data. By doing this, we should be able to use this step to represent $p_\theta(x_{t-1}|x_t)$, and thus recover an image.

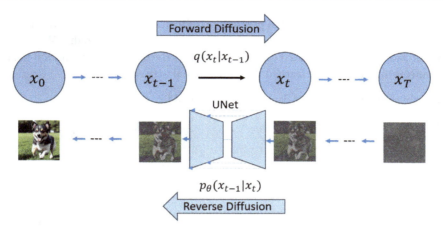

Figure 4.6: Forward diffusion and reverse process

You may ask how we should calculate the loss and update the weights. The ending image (x_T) removes the previously added noise and will provide the ground truth data. After all, we can generate the noise data in the forward diffusion processes on the fly. Next, compare it with the output data from the neural network (usually a UNet). We get the loss data that can be used to calculate the gradient descendant data and update the neural network weights.

The DDPM paper [4] provided a simplified way to calculate the loss:

$$L_{simple}(\theta) := \mathbb{E}_{t, x_0, \in}\left[\left|\left|\ \in\ -\epsilon_\theta\big(\sqrt{\overline{\alpha}_t}x_0 + \sqrt{1 - \overline{\alpha}_t}\,\epsilon, t\big)\ \right|\right|^2\right]$$

Since $x_t = \sqrt{\overline{\alpha}_t x_0} + \sqrt{1 - \overline{\alpha}_t}$, we can further simplify the formula to the following:

$$L_{simple}(\theta) := \mathbb{E}_{t, x_0, \epsilon}\left[\left|\left|\epsilon - \epsilon_\theta(x_t, t)\right|\right|^2\right]$$

The UNet will take a noised image data: x_t and a time step data: t as inputs as shown in *Figure 4.7*. Why take t as input? Because all the denoising processes share the same neural network weights, the input t will help train a UNet with a time step in mind.

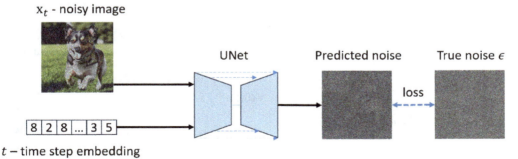

Figure 4.7: UNet training inputs and loss calculation

When we say let's train a neural network to predict the noise distribution that will be removed from the image leading to a clearer image, what is the neural network predicting? In the DDPM paper [4], the original diffusion model uses a fixed variance θ, and sets the Gaussian distribution mean - μ as the only parameter that needs to be learned through a neural network.

In a PyTorch implementation, the loss data can be calculated like this:

```
import torch
import torch.nn as nn

# code prepare the model object, image and timestep
# ...

# noise is the Ɛ ~ N(0,1) with the shape of the image x_t.
noise = torch.randn_like(x_t)
# x_t is the noised image at step "t", together with the time_step value
predicted_noise = model(x_t, time_step)
loss = nn.MSELoss(noise, predicted_noise)

# backward weight prcpagation
# ...
```

Now, we should be able to train a diffusion model and the model should be able to recover an image from a random Gaussian distributed noise. Next, let's take a look at how the inference or sampling works.

The noise-to-image sampling process

Here are the steps to sample an image from the model, or, in other words, generate an image from the reverse diffusion process:

1. Generate a complete Gaussian noise with a mean of 0 and a variance of 1:

$$x_T \sim \mathcal{N}(0,1)$$

    We will use this noise as the starting image.

2. Loop through $t = T$ to $t = 1$. In each step, if $t > 1$, then generate another Gaussian noise image z:

$$z \sim \mathcal{N}(0,1)$$

    If $t = 1$, then the following occurs:

$$z = 0$$

Then, generate a noise from the UNet model, and remove the generated noise from the input noisy image x_t:

$$x_{t-1} = \frac{1}{\sqrt{\alpha_t}}\left(x_t - \frac{1-\alpha_t}{\sqrt{1-\bar{\alpha}_t}}\epsilon_\theta(x_t, t)\right) + \sqrt{1-\alpha_t}\,z$$

If we take a look at the preceding equation, all those α_t and $\bar{\alpha}_t$ are known numbers sourced from β_t. The only thing we need from the UNet is the $\epsilon_\theta(x_t, t)$, which is the noise produced by the UNet, as shown in *Figure 4.8*.

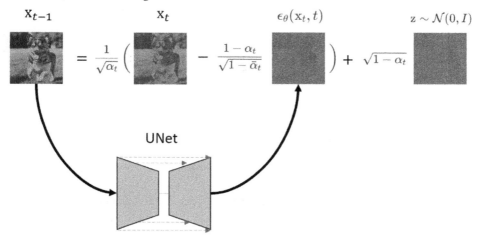

Figure 4.8: Sampling from UNet

The added $\sqrt{1-\alpha_t}\,z$ looks a little bit mysterious here. Why add this to the process? The original paper doesn't explain this added noise, but researchers found that the added noise in the denoising process will significantly improve the generated image quality!

3. Loop end, return the final generated image x_0.

Now, let's talk about the image generation guidance.

Understanding Classifier Guidance denoising

Until now, we haven't talked about the text guidance yet. The image generation process will take a random Gaussian noise as the only input, and then randomly generate an image based on the training dataset. But we want a guided image generation; for example, input "dog" to ask the diffusion model to generate an image including "dog."

In 2021, Dhariwal and Nichol, from OpenAI, proposed classifier guidance in their paper titled *Diffusion Models Beat GANs on Image Synthesis* [12].

Based on the proposed methodology, we can achieve classifier-guided denoising by providing a classification label during the training stage. Instead of just image or time-step embedding, we also provide text description embeddings as shown in *Figure 4.9*.

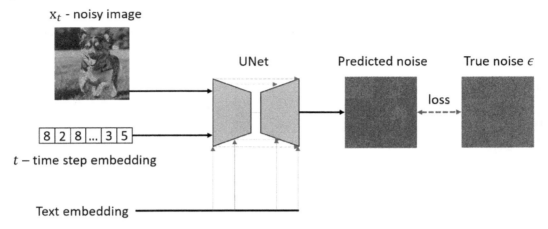

Figure 4.9: Train a diffusion model with conditional text

In *Figure 4.7*, there are two inputs, while in *Figure 4.9*, there is one additional input – **Text embedding**; it is the embedding data generated from OpenAI's CLIP model. We will discuss the way more powerful CLIP model guided diffusion model in the next chapter.

Summary

In this chapter, we took a deep dive into the internal workings of the diffusion model initially brought out by Jonathan Ho et al. [4]. We learned about the foundational ideas of the diffusion model and learned about the forward diffusion process. We also walked through the reverse diffusion process for diffusion model training and sampling and explored how to enable a text-guided diffusion model.

Through this chapter, we aimed to explain the core idea of the diffusion model. If you want to implement a diffusion model by yourself, I would recommend reading through the original DDPM paper directly.

The DDPM diffusion model can generate realistic images, but one of its problems is its performance. Not only is training a model slow, but the image sampling is also slow. In the next chapter, we are going to discuss the Stable Diffusion model, which will boost the speed in a genius way.

References

1. *The Annotated Diffusion Model* – https://colab.research.google.com/github/huggingface/notebooks/blob/main/examples/annotated_diffusion.ipynb#scrollTo=c5a94671

2. *Training with Diffusers* – https://colab.research.google.com/gist/anton-l/f3a8206dae4125b93f05b1f5f703191d/diffusers_training_example.ipynb

3. *Diffusers* – https://colab.research.google.com/github/huggingface/notebooks/blob/main/diffusers/diffusers_intro.ipynb#scrollTo=PzW5ublpBuUt

4. Jonathan Ho et al., *Denoising Diffusion Probabilistic Models* – https://arxiv.org/abs/2006.11239

5. Steins, *Diffusion Model Clearly Explained!* – https://medium.com/@steinsfu/diffusion-model-clearly-explained-cd331bd41166

6. Steins, *Stable Diffusion Clearly Explained!* – https://medium.com/@steinsfu/stable-diffusion-clearly-explained-ed008044e07e

7. DeepFindr, *Diffusion models from scratch in PyTorch* – https://www.youtube.com/watch?v=a4Yfz2FxXiY&t=5s&ab_channel=DeepFindr

8. Ari Seff, *What are Diffusion Models?* – https://www.youtube.com/watch?v=fbLgFrlTnGU&ab_channel=AriSeff

9. Prafulla Dhariwal, Alex Nichol, *Diffusion Models Beat GANs on Image Synthesis* – https://arxiv.org/abs/2105.05233

10. Diederik P Kingma, Max Welling, *Auto-Encoding Variational Bayes* – https://arxiv.org/abs/1312.6114

11. Lilian Weng, *What are Diffusion Models?* – https://lilianweng.github.io/posts/2021-07-11-diffusion-models/

12. Prafulla Dhariwal, Alex Nichol, *Diffusion Models Beat GANs on Image Synthesis* – https://arxiv.org/abs/2105.05233

5
Understanding How Stable Diffusion Works

In *Chapter 4*, we dove into the internal workings of the diffusion model with some math formulas. If you are not used to reading the formulas every day, it can be scary, but once you get familiar with those symbols and Greek letters, the benefit of fully understanding those formulas is huge. Math formulas and equations not only help us understand the core of the process in a precise and concise form, but they also enable us to read more papers and works from others.

While the original diffusion model is more like a proof of a concept, it shows the huge potential of the multi-step diffusion model compared with a one-pass neural network. However, some drawbacks come with the original diffusion model, **denoising diffusion probabilistic models (DDPM)** [1], and later Classifier Guidance denoising. Let me list two:

- To train a diffusion model with Classifier Guidance requires training a new classifier, and we can't reuse a pre-trained classifier. Also, in diffusion model training, training a classifier with 1,000 categories is already not easy.

- Pre-trained model inferences in pixel space are computationally expensive, not to mention training a model. Using a pre-trained model to generate 512x512 images in pixel space on a home computer with 8 GB of VRAM, without memory optimization, is not possible.

In 2022, researchers proposed **Latent Diffusion models**, Robin et al [2]. The model nicely solved both the classification problem and the performance problem. The Latent Diffusion model was later known as Stable Diffusion.

In this chapter, we will take a look at how Stable Diffusion solved the preceding problems and led to state-of-the-art developments in the field of image generation. We will specifically cover the following topics:

- Stable Diffusion in latent space
- Generating latent vectors using Diffusers
- Generating text embeddings using CLIP
- Generating time step embeddings
- Initializing Stable Diffusion UNet
- Implementing a text-to-image Stable Diffusion inference pipeline
- Implementing a text-guided image-to-image Stable Diffusion inference pipeline
- Putting all the code together

Let's dive into the core of Stable Diffusion.

The sample code for this chapter is tested using version 0.20.0 of the Diffusers package. To ensure the code runs smoothly, please use Diffusers v0.20.0. You can install it using the following command:

```
pip install diffusers==0.20.0
```

Stable Diffusion in latent space

Instead of processing diffusion in pixel space, Stable Diffusion uses latent space to represent an image. What is latent space? In short, latent space is the vector representation of an object. To use an analogy, before you go on a blind date, a matchmaker could provide you with your counterpart's height, weight, age, hobbies, and so on in the form of a vector:

```
[height, weight, age, hobbies,...]
```

You can take this vector as the latent space of your blind date counterpart. A real person's true property dimension is almost unlimited (you could write a biography for one). The latent space can be used to represent a real person with only a limited number of features, such as height, weight, and age.

In the case of the Stable Diffusion training stage, a trained encoder model, usually denoted as $\mathcal{E}$ *(E)*, is used to encode an input image in a latent vector representation. After the reverse diffusion process, the latent space is decoded by a decoder in pixel space. The decoder is usually denoted as *D (D)*.

Both training and sampling work take place in the latent space. The training process is shown in *Figure 5.1*:

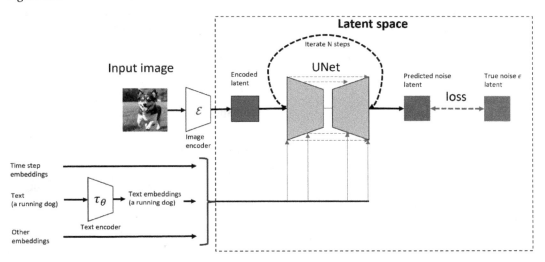

Figure 5.1: Training Stable Diffusion model in latent space

Figure 5.1 illustrates the training process of the Stable Diffusion model. It shows a high-level overview of how the model is trained.

Here is a step-by-step breakdown of the process:

1. **Inputs**: The model is trained using images, caption text, and time step embeddings (specifying at which step the denoising happens).

2. **Image encoder**: The input image is passed through an encoder. The encoder is a neural network that processes the input image and converts it into a more abstract and compressed representation. This representation is often referred to as a "latent space" because it captures the image's underlying characteristics, but not the pixel-level details.

3. **Latent space**: The encoder outputs a vector that represents the input image in the latent space. The latent space is typically a lower-dimensional space than the input space (the pixel space of the image), which allows for faster processing and more efficient representation of the input data. The whole training happens in the latent space.

4. **Iterate N steps**: The training process involves iterating through the latent space multiple times (N steps). This iterative process is where the model learns to refine the latent space representation and make small adjustments to match the desired output image.

5. **UNet**: After each iteration, the model uses UNet to generate an output image based on the current latent space vector. UNet generates the predicted noise and incorporates the input text embedding, step information, and potentially other embeddings.

6. **The loss function**: The model's training process also involves a loss function. This measures the difference between the output image and the desired output image. As the model iterates, the loss is continually calculated, and the model makes adjustments to its weights to minimize this loss. This is how the model learns from its mistakes and improves over time.

Refer to *Chapter 21* for more detailed steps on model training.

The process of inferencing from UNet is shown in *Figure 5.2*:

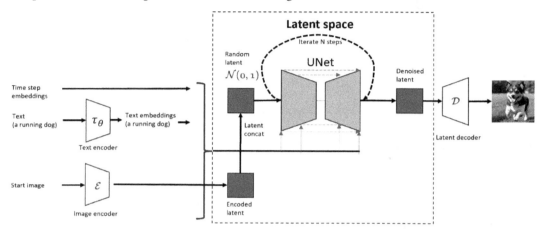

Figure 5.2: Stable Diffusion inferencing in latent space

Stable Diffusion not only supports text-guided image generation; it also supports image-guided generation.

In *Figure 5.2*, starting from the left side, we can see that both text and an image are used to guide the image generation.

When we provide a text input, Stable Diffusion uses CLIP [3] to generate an embedding vector, which will be fed into UNet, using the attention mechanism.

When we provide an image as the guiding signal, the input image will be encoded to latent space and then concatenate with the randomly generated Gaussian noise.

It is all up to us to provide guidance; we can provide either text, an image, or both. We can even generate images without providing any images; in this "empty" guidance case, the UNet model will decide what to generate based on the randomly initialized noise.

With the two essential inputs provided text embeddings and the initial image latent noise (with or without the initial image's encoded vectors in latent space), UNet kicks off to remove noise from the initial image in the latent space. After several denoising steps, with the help of a decoder, Stable Diffusion can output a vivid image in pixel space.

The process is similar to the training process but without sending the loss value back to update the weights. Instead, after a number of denoising steps (*N* steps), the latent decoder (**Variational Autoencoder (VAE)** [4]) converts the image from latent space to visible pixel space.

Next, let's take a look at what those components (the text encoder, image Encoder, UNet, and image decoder) look like, and then we'll build one of our own Stable Diffusion pipelines step by step.

Generating latent vectors using diffusers

In this section, we are going to use a pre-trained Stable Diffusion model to encode an image into latent space so that we have a concrete impression of what a latent vector looks and feels like. Then, we will decode the latent vector back into an image. This operation will also establish the foundation for building the image-to-image custom pipeline:

1. **Load an image**: We can use the `load_image` function from `diffusers` to load an image from local storage or a URL. In the following code, we load an image named `dog.png` from the same directory of the current program:

```
from diffusers.utils import load_image
image = load_image("dog.png")
display(image)
```

2. **Pre-process the image**: Each pixel of the loaded image is represented by a number ranging from 0 to 255. The image encoder from the Stable Diffusion process handles image data ranging from -1.0 to 1.0. So, we first need to make the data range conversion:

```
import numpy as np

# convert image object to array and
# convert pixel data from 0 ~ 255 to 0 ~ 1
image_array = np.array(image).astype(np.float32)/255.0

# convert the number from 0 ~ 1 to -1 ~ 1
image_array = image_array * 2.0 - 1.0
```

Now, if we use Python code, `image_array.shape`, to check the `image_array` data shape, we will see the shape of the image data as – `(512,512,3)`, arranged as `(width, height, channel)`, instead of the commonly used `(channel, width, height)`. Here, we need to convert the image data shape to `(channel, width, height)` or `(3,512,512)`, using the `transpose()` function:

```
# transform the image array from width,height,
# channel to channel,width,height
image_array_cwh = image_array.transpose(2,0,1)
```

The 2 is in the first position of 2, 0, 1, which means moving the original third dimension (indexed as 2) to the first dimension. The same logic applies to 0 and 1. The original 0 dimension is now converted to the second position, and the original 1 is now in the third dimension.

With this transpose operation, the NumPy array, image_array_cwh, is now in the (3,512,512) shape.

The Stable Diffusion image encoder handles image data in batches, which, in this instance is four-dimensional data with the batch dimension in the first position; we need to add the batch dimension here:

```
# add batch dimension
image_array_cwh = np.expand_dims(image_array_cwh, axis = 0)
```

3. **Load image data with** torch **and move to CUDA**: We will convert the image data to latent space using CUDA. To achieve this, we will need to load the data into the CUDA VRAM before handing it off to the next step model:

```
# load image with torch
import torch
image_array_cwh = torch.from_numpy(image_array_cwh)
image_array_cwh_cuda = image_array_cwh.to(
    "cuda",
    dtype=torch.float16
)
```

4. **Load the Stable Diffusion image encoder VAE**: This VAE model is used to convert the image from pixel space to latent space:

```
# Initialize VAE model
import torch
from diffusers import AutoencoderKL

vae_model = AutoencoderKL.from_pretrained(
    "runwayml/stable-diffusion-v1-5",
    subfolder = "vae",
    torch_dtype=torch.float16
).to("cuda")
```

5. **Encode the image into a latent vector**: Now, everything is ready, and we can encode any image into a latent vector as PyTorch tensor:

```
latents = vae_model.encode(
    image_array_cwh_cuda).latent_dist.sample()
```

Check the data and shape of the latent data:

```
print(latents[0])
print(latents[0].shape)
```

We can see that the latent is in the `(4, 64, 64)` shape, with each element in the range of `-1.0` to `1.0`.

Stable Diffusion processes all the denoising steps on a 64x64 tensor with 4-channel for a 512x512 image generation. The data size is way less than its original image size, 512x512 with three color channels.

6. **Decode latent to image (optional)**: You may be wondering, can I convert the latent data back to the pixel image? Yes, we can do this with lines of code:

```
import numpy as np
from PIL import Image

def latent_to_img(latents_input, scale_rate = 1):
    latents_2 = (1 / scale_rate) * latents_input

    # decode image
    with torch.no_grad():
        decode_image = vae_model.decode(
        latents_input,
        return_dict = False
        )[0][0]

    decode_image =  (decode_image / 2 + 0.5).clamp(0, 1)

    # move latent data from cuda to cpu
    decode_image = decode_image.to("cpu")

    # convert torch tensor to numpy array
    numpy_img = decode_image.detach().numpy()

    # covert image array from (width, height, channel)
    # to (channel, width, height)
    numpy_img_t = numpy_img.transpose(1,2,0)

    # map image data to 0, 255, and convert to int number
    numpy_img_t_01_255 = \
        (numpy_img_t*255).round().astype("uint8")

    # shape the pillow image object from the numpy array
    return Image.fromarray(numpy_img_t_01_255)
```

```
pil_img = latent_to_img(latents_input)
pil_img
```

The diffusers Stable Diffusion pipeline will finally generate a latent tensor. We will follow similar steps to recover a denoised latent for an image in the latter part of this chapter.

Generating text embeddings using CLIP

To generate the text embeddings (the embeddings contain the image features), we need first to tokenize the input text or prompt and then encode the token IDs into embeddings. Here are steps to achieve this:

1. **Get the prompt token IDs**:

    ```
    input_prompt = "a running dog"

    # input tokenizer and clip embedding model
    import torch
    from transformers import CLIPTokenizer,CLIPTextModel

    # initialize tokenizer
    clip_tokenizer = CLIPTokenizer.from_pretrained(
        "runwayml/stable-diffusion-v1-5",
        subfolder = "tokenizer",
        dtype     = torch.float16
    )
    input_tokens = clip_tokenizer(
        input_prompt,
        return_tensors = "pt"
    )["input_ids"]
    input_tokens
    ```

    The preceding code will convert the a running dog text prompt to a token ID list as a torch tensor object – tensor([[49406,   320, 2761, 1929, 49407]]).

2. **Encode the token IDs into embeddings**:

    ```
    # initialize CLIP text encoder model
    clip_text_encoder = CLIPTextModel.from_pretrained(
        "runwayml/stable-diffusion-v1-5",
        subfolder="text_encoder",
        # dtype=torch.float16
    ).to("cuda")

    # encode token ids to embeddings
    ```

```
prompt_embeds = clip_text_encoder(
    input_tokens.to("cuda")
)[0]
```

3. **Check the embedding data**:

```
print(prompt_embeds)
print(prompt_embeds.shape)
```

Now, we can see the data of `prompt_embeds` as follows:

```
tensor([[[-0.3884,0.0229, -0.0522,..., -0.4899, -0.3066,0.0675],
    [ 0.0290, -1.3258,  0.3085,..., -0.5257,0.9768,0.6652],
    [ 1.4642,0.2696,0.7703,..., -1.7454, -0.3677,0.5046],
    [-1.2369,0.4149,1.6844,..., -2.8617, -1.3217,0.3220],
    [-1.0182,0.7156,0.4969,..., -1.4992, -1.1128, -0.2895]]],
    device='cuda:0', grad_fn=<NativeLayerNormBackward0>)
```

Its shape is `torch.Size([1, 5, 768])`. Each token ID is encoded into a 768-dimension vector.

4. **Generate embedding for negative prompt embeddings**: Even though we don't have the negative prompt, we'll also prepare an embedding vector with the same size as the input prompt. This will ensure that our code will support both only `prompt` and `prompt/negative prompt` cases:

```
# prepare neg prompt embeddings
uncond_tokens = "blur"

# get the prompt embedding length
max_length = prompt_embeds.shape[1]

# generate negative prompt tokens with the same length of prompt
uncond_input_tokens = clip_tokenizer(
    uncond_tokens,
    padding = "max_length",
    max_length = max_length,
    truncation = True,
    return_tensors = "pt"
)["input_ids"]

# generate the negative embeddings
with torch.no_grad():
    negative_prompt_embeds = clip_text_encoder(
        uncond_input_tokens.to("cuda")
    )[0]
```

5. **Concatenate prompt and negative prompt embedding into one vector**: Because we will feed the whole prompt into UNet at once, and then handle the positive and negative signals at the UNet inference stage, we will concatenate the prompt and negative prompt embeddings into one `torch` vector:

```
prompt_embeds = torch.cat([negative_prompt_embeds,
    prompt_embeds])
```

Next, we will initialize the time step data.

Initializing time step embeddings

We introduced the scheduler in *Chapter 3*. By using the scheduler, we can sample key steps for image generation. Instead of denoising 1,000 steps to generate an image in the original diffusion model (DDPM), by using a scheduler, we can generate an image in a mere 20 steps.

In this section, we are going to use the Euler scheduler to generate time step embeddings, and then we'll take a look at what the time step embeddings look like. No matter how good the diagram that tries to plot the process is, we can only understand how it works by reading the actual data and code:

1. **Initialize a scheduler from the scheduler configuration for the model**:

```
from diffusers import EulerDiscreteScheduler as Euler

# initialize scheduler from a pretrained checkpoint
scheduler = Euler.from_pretrained(
    "runwayml/stable-diffusion-v1-5",
    subfolder = "scheduler"
)
```

The preceding code will initialize a scheduler from the checkpoint's scheduler config file. Note that you can also create a scheduler, as we discussed in *Chapter 3*, like this:

```
import torch
from diffusers import StableDiffusionPipeline
from diffusers import EulerDiscreteScheduler as Euler

text2img_pipe = StableDiffusionPipeline.from_pretrained(
    "runwayml/stable-diffusion-v1-5",
    torch_dtype = torch.float16
).to("cuda:0")

scheduler = Euler.from_config(text2img_pipe.scheduler.config)
```

However, this will require you to load a model first, which is not only slow but also unnecessary; the only thing we need is the model's scheduler.

2. **Sample the steps for the image diffusion process**:

```
inference_steps = 20
scheduler.set_timesteps(inference_steps, device = "cuda")

timesteps = scheduler.timesteps
for t in timesteps:
    print(t)
```

We will see the 20-step value as follows:

```
...
tensor(999., device='cuda:0', dtype=torch.float64)
tensor(946.4211, device='cuda:0', dtype=torch.float64)
tensor(893.8421, device='cuda:0', dtype=torch.float64)
tensor(841.2632, device='cuda:0', dtype=torch.float64)
tensor(788.6842, device='cuda:0', dtype=torch.float64)
tensor(736.1053, device='cuda:0', dtype=torch.float64)
tensor(683.5263, device='cuda:0', dtype=torch.float64)
tensor(630.9474, device='cuda:0', dtype=torch.float64)
tensor(578.3684, device='cuda:0', dtype=torch.float64)
tensor(525.7895, device='cuda:0', dtype=torch.float64)
tensor(473.2105, device='cuda:0', dtype=torch.float64)
tensor(420.6316, device='cuda:0', dtype=torch.float64)
tensor(368.0526, device='cuda:0', dtype=torch.float64)
tensor(315.4737, device='cuda:0', dtype=torch.float64)
tensor(262.8947, device='cuda:0', dtype=torch.float64)
tensor(210.3158, device='cuda:0', dtype=torch.float64)
tensor(157.7368, device='cuda:0', dtype=torch.float64)
tensor(105.1579, device='cuda:0', dtype=torch.float64)
tensor(52.5789, device='cuda:0', dtype=torch.float64)
tensor(0., device='cuda:0', dtype=torch.float64)
```

Here, the scheduler takes 20 steps out of the 1,000 steps, and those 20 steps may be enough to denoise a complete Gaussian distribution for image generation. This step sampling technique also contributes to Stable Diffusion performance boosting.

Initializing the Stable Diffusion UNet

The UNet architecture [5] was introduced by Ronneberger et al. for biomedical image segmentation purposes. Before the UNet architecture, a convolution network was commonly used for image classification tasks. When using a convolution network, the output is a single class label. However, in many visual tasks, the desired output should include localization too, and the UNet model solved this problem.

The U-shaped architecture of UNet enables efficient learning of features at different scales. UNet's skip connections directly combine feature maps from different stages, allowing a model to effectively propagate information across various scales. This is crucial for denoising, as it ensures the model retains both fine-grained details and global context during noise removal. These features make UNet a good candidate for the denoising model.

In the `Diffuser` library, there is a class named `UNet2DconditionalModel`; this is a conditional 2D UNet model for image generation and related tasks. It is a key component of diffusion models and plays a crucial role in the image generation process. We can load a UNet model in just several lines of code, like this:

```
import torch
from diffusers import UNet2DConditionModel

unet = UNet2DConditionModel.from_pretrained(
    "runwayml/stable-diffusion-v1-5",
    subfolder ="unet",
    torch_dtype = torch.float16
).to("cuda")
```

Together with the UNet model we have just loaded up, we have all the components required by Stable Diffusion. Not that hard, right? Next, we are going to use those building blocks to build two Stable Diffusion pipelines – one text-to-image and another image-to-image.

Implementing a text-to-image Stable Diffusion inference pipeline

So far, we have all the text encoder, image VAE, and denoising UNet model initialized and loaded into the CUDA VRAM. The following steps will chain them together to form the simplest and working Stable Diffusion text-to-image pipeline:

1. **Initialize a latent noise:** In *Figure 5.2*, the starting point of inference is randomly initialized Gaussian latent noise. We can create one of the latent noise with this code:

```
# prepare noise latents
shape = torch.Size([1, 4, 64, 64])
device = "cuda"
noise_tensor = torch.randn(
    shape,
    generator = None,
    dtype      = torch.float16
).to("cuda")
```

During the training stage, an initial noise sigma is used to help prevent the diffusion process from becoming stuck in local minima. When the diffusion process starts, it is very likely to be in a state where it is very close to a local minimum. `init_noise_sigma = 14.6146` is used to help avoid this. So, during the inference, we will also use `init_noise_sigma` to shape the initial latent.

```
# scale the initial noise by the standard deviation required by
# the scheduler
latents = noise_tensor * scheduler.init_noise_sigma
```

2. **Loop through UNet**: With all those components prepared, we are finally at the stage of feeding the initial latents to UNet to generate the target latent we want:

```
guidance_scale = 7.5
latents_sd = torch.clone(latents)
for i,t in enumerate(timesteps):
    # expand the latents if we are doing classifier free guidance
    latent_model_input = torch.cat([latents_sd] * 2)
    latent_model_input = scheduler.scale_model_input(
        latent_model_input, t)

    # predict the noise residual
    with torch.no_grad():
        noise_pred = unet(
            latent_model_input,
            t,
            encoder_hidden_states=prompt_embeds,
            return_dict = False,
        )[0]

    # perform guidance
    noise_pred_uncond, noise_pred_text = noise_pred.chunk(2)
    noise_pred = noise_pred_uncond + guidance_scale *
        (noise_pred_text - noise_pred_uncond)

    # compute the previous noisy sample x_t -> x_t-1
    latents_sd = scheduler.step(noise_pred, t,
        latents_sd, return_dict=False)[0]
```

The preceding code is a simplified denoising loop of `DiffusionPipeline` from the `diffusers` package, removing all those edging cases and only keeping the core of the inferencing.

The algorithm works by iteratively adding noise to a latent representation of an image. In each iteration, the noise is guided by a text prompt, which helps the model generate images that are more similar to the prompt.

The preceding code first defines a few variables:

- The `guidance_scale` variable controls the amount of guidance that is applied to the noise.

- The `latents_sd` variable stores the latent representation of the image that is generated. The time steps variable stores a list of time steps at which the noise will be added.

The main loop of the code iterates over the time steps. In each iteration, the code first expands the latent representation to include two copies of itself. This is done because the Stable Diffusion algorithm uses a classifier-free guidance mechanism, which requires two copies of the latent representation.

The code then calls the `unet` function to predict the noise residual for the current time step.

The code then performs guidance on the noise residual. This involves adding a scaled version of the text-conditioned noise residual to the unconditional noise residual. The amount of guidance that is applied is controlled by the `guidance_scale` variable.

Finally, the code calls the `scheduler` function to update the latent representation of the image. The `scheduler` function is a function that controls the amount of noise that is added to the latent representation at each time step.

As mentioned previously, the preceding code is a simplified version of the Stable Diffusion algorithm. In practice, the algorithm is much more complex, and it incorporates a number of other techniques to improve the quality of the generated images.

3. **Recover the image from the latent**: We can reuse the `latent_to_img` function to recover the image from the latent space:

```python
import numpy as np
from PIL import Image

def latent_to_img(latents_input):
    # decode image
    with torch.no_grad():
        decode_image = vae_model.decode(
            latents_input,
            return_dict = False
        )[0][0]

    decode_image = (decode_image / 2 + 0.5).clamp(0, 1)

    # move latent data from cuda to cpu
    decode_image = decode_image.to("cpu")
```

```
    # convert torch tensor to numpy array
    numpy_img = decode_image.detach().numpy()

    # covert image array from (channel, width, height)
    # to (width, height, channel)
    numpy_img_t = numpy_img.transpose(1,2,0)

    # map image data to 0, 255, and convert to int number
    numpy_img_t_01_255 = \
        (numpy_img_t*255).round().astype("uint8")

    # shape the pillow image object from the numpy array
    return Image.fromarray(numpy_img_t_01_255)

latents_2 = (1 / 0.18215) * latents_sd
pil_img = latent_to_img(latents_2)
```

The `latent_to_img` function performs actions in the following sequence:

I. It calls the `vae_model.decode` function to decode the latent vector into an image. The `vae_model.decode` function is a function that is trained on a dataset of images. It can be used to generate new images that are similar to the images in the dataset.

II. Normalizes the image data to a range of 0 to 1. This is done because the `Image.fromarray` function expects image data to be in this range.

III. Moves the image data from the GPU to the CPU. Then, it converts the image data from a torch tensor to a NumPy array. This is done because the `Image.fromarray` function only accepts NumPy arrays as input.

IV. Flips the dimensions of the image array so that it is in the (width, height, channel) format, the format that the `Image.fromarray` function expects.

V. Maps the image data to a range from 0 to 255 and converts it to an integer type.

VI. Calls the `Image.fromarray` function to create a Python imaging library (PIL) image object from the image data.

The `latents_2 = (1 / 0.18215) * latents_sd` line of code is needed when decoding the latent to image because the latents are scaled by a factor of 0.18215 during training. This scaling is done to ensure that latent space has a unit variance. When decoding, the latents need to be scaled back to their original scale to reconstruct the original image.

Then, you should be able to see something like this if everything is going well:

Figure 5.3: A running dog, generated by a custom Stable Diffusion pipeline

In the next section, we are going to implement an image-to-image Stable Diffusion pipeline.

Implementing a text-guided image-to-image Stable Diffusion inference pipeline

The only thing we need to do now is concatenate the starting image with the starting latent noise. The `latents_input` Torch tensor is the latent we encoded from a dog image earlier in this chapter:

```
strength = 0.7
# scale the initial noise by the standard deviation required by the
# scheduler
latents = latents_input*(1-strength) +
    noise_tensor*scheduler.init_noise_sigma
```

That is all that is necessary; use the same code from the text-to-image pipeline, and you should generate something like *Figure 5.4*:

Figure 5.4: A running dog, generated by a custom image-to-image Stable Diffusion pipeline

Note that the preceding code uses `strength = 0.7`; the strength denotes the weight of the original latent noise. If you want an image more similar to the initial image (the image you provided to the image-to-image pipeline), use a lower strength number; otherwise, increase it.

Summary

In this chapter, we moved on from the original diffusion model, DDPM, and explained what Stable Diffusion is and why it is faster and better than the DDPM model.

As suggested by the paper *High-Resolution Image Synthesis with Latent Diffusion Models* [6] that introduced Stable Diffusion, the biggest feature that differentiates Stable Diffusion from its predecessor is the "*Latent.*" This chapter explained what latent space is and how Stable Diffusion training and inference work internally.

For a comprehensive understanding, we created components using methods such as encoding the initial image into latent data, converting input prompts to token IDs and embedding them to text embeddings using the CLIP text model, using the Stable Diffusion scheduler to sample detailed steps for inference, creating the initial noise latent, concatenating initial noise latent with the initial image latent, putting all the components together to build a custom text-to-image Stable Diffusion pipeline, and extending the pipeline to enable a text-guided image-to-image Stable Diffusion pipeline. We created these components one by one, and finally, we built two Stable Diffusion pipelines – one text-to-image pipeline and an extended text-guided image-to-image pipeline.

By completing this chapter, you should not only have a general understanding of Stable Diffusion but also be able to freely build your own pipelines to meet specific requirements.

In the next chapter, we are going to introduce solutions to load Stable Diffusion models.

References

1. Jonathan Ho, Ajay Jain, Pieter Abbeel, Denoising Diffusion Probabilistic Models: `https://arxiv.org/abs/2006.11239`

2. Robin et al, High-Resolution Image Synthesis with Latent Diffusion Models: `https://arxiv.org/abs/2112.10752`

3. Alec et al, Learning Transferable Visual Models From Natural Language Supervision: `https://arxiv.org/abs/2103.00020`

4. VAEs: `https://en.wikipedia.org/wiki/Variational_autoencoder`

5. UNet2DConditionModel document from Hugging Face: `https://huggingface.co/docs/diffusers/api/models/unet2d-cond`

6. Robin et al, High-Resolution Image Synthesis with Latent Diffusion Models: `https://arxiv.org/abs/2112.10752`

Additional reading

Jonathan Ho, Tim Salimans, Classifier-Free Diffusion Guidance: `https://arxiv.org/abs/2207.12598`

Stable Diffusion with Diffusers: `https://huggingface.co/blog/stable_diffusion`

Olaf Ronneberger, Philipp Fischer, Thomas Brox, UNet: Convolutional Networks for Biomedical Image Segmentation: `https://arxiv.org/abs/1505.04597`

6

Using Stable Diffusion Models

When we start using Stable Diffusion models, we will immediately encounter different kinds of model files and will need to know how to convert a model file to the desired format.

In this chapter, we are going to get more familiar with Stable Diffusion model files, covering how to load models from the Hugging Face repository using model IDs. We'll also provide sample code to load `safetensors` and `.ckpt` model files shared by the open source community.

In this chapter, we will cover the following topics:

- Loading the Diffusers model
- Loading model checkpoints from safetensors and ckpt files
- Using CKPT and safetensors files with Diffusers
- Model safety checker
- Converting checkpoint model files to the Diffusers format
- Using Stable Diffusion XL

By the end of this chapter, you will have learned about the Stable Diffusion model file types and how to convert and load model files to a format that can be loaded with Diffusers.

Technical requirements

Before you start, make sure you have the `safetensors` package installed:

```
pip install safetensors
```

The `safetensors` Python package offers a simple and efficient way to access, store, and share tensors securely.

Loading the Diffusers model

Instead of downloading model files manually, the Hugging Face Diffusers package provides a convenient way to access open source model files from a string-type model ID like this:

```
import torch
from diffusers import StableDiffusionPipeline
pipe = StableDiffusionPipeline.from_pretrained(
    "runwayml/stable-diffusion-v1-5",
    torch_dtype = torch.float16
)
```

When the preceding code is executed, if Diffusers can't find the model files that are denoted by the model ID, the package will automatically reach out to the Hugging Face repository to download the model files and store them in a cache folder for next time.

By default, the cache files will be stored in the following places:

Windows:

```
C:\Users\user_name\.cache\huggingface\hub
```

Linux:

```
\home\user_name\.cache\huggingface\hub
```

Using the default cache path is fine in the beginning, however, if your system driver is less than 512 GB, you will soon find those model files are eating up storage space. To avoid running out of storage, we may need to plan the model storage in advance. Diffusers provides a parameter for us to specify a custom path for storing the cached weight files.

The following is the preceding sample code with one more parameter, cache_dir:

```
from diffusers import StableDiffusionPipeline
pipe = StableDiffusionPipeline.from_pretrained(
    "runwayml/stable-diffusion-v1-5",
    torch_dtype = torch.float16,
    cache_dir = r"D:\my_model_folder"
)
```

By specifying this cache_dir parameter, all auto-downloaded model and configuration files will be stored in the new location instead of eating up the system disk drive.

You might also notice that the sample code specifies a torch_dtytpe parameter to tell Diffusers to use torch.float16. By default, PyTorch uses torch.float32 for matrix multiplications. For model inference, or in other words, at the stage of using Stable Diffusion to generate images, we can use the float16 type to not only increase the speed by about 100% but also save GPU memory with almost unnoticeable difference.

Usually, using models from Hugging Face is easy and safe. Hugging Face implements a safety checker to ensure the uploaded model files do not contain any malicious code that may harm your computer.

Nevertheless, we can still use manually downloaded model files with Diffusers. Next, we are going to load various model files from the local disk.

Loading model checkpoints from safetensors and ckpt files

The complete model files are also called **checkpoint** data. If you read an article or document talking about downloading a checkpoint, they are talking about a Stable Diffusion model file.

There are many types of checkpoints, such as .ckpt files, safetensors files, and diffusers files:

- .ckpt is the most basic file format and is compatible with most Stable Diffusion models. However, they are also the most vulnerable to malicious attacks.

- safetensors is a newer file format that is designed to be more secure than .ckpt files. The safetensors format is better in terms of security, speed, and usability compared with .ckpt. Safetensors has several features to prevent code execution:

  - **Restricted data types**: Only specific data types, such as integers and tensors, are allowed to be stored. This eliminates the possibility of including code within the saved data.

  - **Hashing**: Each chunk of data is hashed, and the hash is stored alongside the data. Any modification to the data would change the hash, making it instantly detectable.

  - **Isolation**: Data is stored in an isolated environment, preventing interaction with other programs, and protecting your system from potential exploits.

- Diffusers files are the latest file format specifically crafted for seamless integration with the Diffusers library. This format boasts top-notch security features and ensures compatibility with all Stable Diffusion models. Unlike traditional compression into a single file, the Diffusers format takes the form of a folder that encompasses both weights and configuration files. Moreover, the model files contained within these folders adhere to the safetensors format.

When we use the Diffusers auto download function, Diffusers will store the files in the Diffusers format.

Next, we are going to load up a Stable Diffusion model in ckpt or safetensors format.

Using ckpt and safetensors files with Diffusers

The Diffusers community is actively enhancing the functionality. At the time of writing, we can easily load .ckpt or safetensors checkpoint files using the Diffusers package.

The following code can be used to load and use a safetensors or .ckpt checkpoint file.

Load the `safetensors` model:

```
import torch
from diffusers import StableDiffusionPipeline
model_path = r"model/path/path/model_name.safetensors"
pipe = StableDiffusionPipeline.from_single_file(
    model_path,
    torch_dtype = torch.float16
)
```

Load the `.ckpt` model with the following code:

```
import torch
from diffusers import StableDiffusionPipeline
model_path = r"model/path/path/model_name.ckpt"
pipe = StableDiffusionPipeline.from_single_file(
    model_path,
    torch_dtype = torch.float16
)
```

You are not reading the wrong code; we can load both `safetensors` and `.ckpt` model files with the same function – `from_single_file`. Next, let's take a look at the safety checker.

Turning off the model safety checker

By default, the Diffusers pipeline will check the output result with a safety checker model to ensure the generated result does not include any NSFW, violent, or unsafe content. In certain cases, the safety checker may trigger false alarms and produce empty images (completely black images). There are several GitHub issue discussions about the safety checker [11]. In the test stage, we can temporarily turn off the safety checker.

To turn off the safety checker when loading the model using the model ID, run the following code:

```
import torch
from diffusers import StableDiffusionPipeline
pipe = StableDiffusionPipeline.from_pretrained(
    "runwayml/stable-diffusion-v1-5",
    torch_dtype    = torch.float16,
    safety_checker = None # or load_safety_checker = False
)
```

Note that the parameter to turn off the safety checker is different when we are loading the model from a `safetensors` or `.ckpt` file. Instead of using `safety_checker`, we should use `load_safety_checker` as shown in the following sample code:

```
import torch
from diffusers import StableDiffusionPipeline
model_path = r"model/path/path/model_name.ckpt"
pipe = StableDiffusionPipeline.from_single_file(
    model_path,
    torch_dtype = torch.float16,
    load_safety_checker = False
)
```

You should be able to use `load_safety_checker = False` in the `from_pretrained` function to disable the safety checker.

The safety checker is an open source machine learning model from CompVis – Computer Vision and Learning LMU Munich (`https://github.com/CompVis`), built based on CLIP [9][10], called **Stable Diffusion Safety Checker** [3].

While we can load a model in a single file, in some cases, we will need to convert a `.ckpt` or `safetensors` model file to the Diffusers folder structure. Next, let's see how we can convert model files to the Diffusers format.

Converting the checkpoint model file to the Diffusers format

Loading checkpoint model data from a `.ckpt` or `safetensors` file is slow compared with the Diffusers format because every time we load a `.ckpt` or `safetensors` file, Diffusers will unpack and convert the file to the Diffusers format. To save the conversion every time we load a model file, we may consider converting checkpoint files to the Diffusers format.

We can use the following code to convert a `.ckpt` file to the Diffusers format:

```
ckpt_checkpoint_path = r"D:\temp\anythingV3_fp16.ckpt"
target_part = r"D:\temp\anythingV3_fp16"
pipe = download_from_original_stable_diffusion_ckpt(
    ckpt_checkpoint_path,
    from_safetensors = False,
    device = "cuda:0"
)
pipe.save_pretrained(target_part)
```

To convert a `safetensors` file to the Diffusers format, simply change the `from_safetensors` parameter to `True` as shown in the following sample code:

```
from diffusers.pipelines.stable_diffusion.convert_from_ckpt import \
    download_from_original_stable_diffusion_ckpt

safetensors_checkpoint_path = \
    r"D:\temp\deliberate_v2.safetensors"
target_part = r"D:\temp\deliberate_v2"
pipe = download_from_original_stable_diffusion_ckpt(
    safetensors_checkpoint_path,
    from_safetensors  = True,
    device = "cuda:0"
)
pipe.save_pretrained(target_part)
```

If you have tried asking a search engine to find a solution to do the conversion, from some corners of the internet, you may see a solution that uses a script called `convert_original_stable_diffusion_to_diffusers.py`. The script is located in the Diffusers GitHub repository: `https://github.com/huggingface/diffusers/tree/main/scripts`. The script works well. If you look at the code of the script, the script uses the same code presented previously.

To use the converted model file, simply use the `from_pretrained` function to load the `local` folder (instead of the model ID) this time:

```
# load local diffusers model files using from_pretrained function
import torch
from diffusers import StableDiffusionPipeline
pipe = StableDiffusionPipeline.from_pretrained(
    r"D:\temp\deliberate_v2",
    torch_dtype = torch.float16,
    safety_checker = None
).to("cuda:0")
image = pipe("a cute puppy").images[0]
image
```

You should see a cute puppy image generated by the preceding code. Next, let's load Stable Diffusion XL models.

Using Stable Diffusion XL

Stable Diffusion XL (SDXL) is a model from Stability AI. Slightly different compared to previous models, SDXL is designed to be a two-stage model. We will need the base model to generate an image and can leverage a second, refiner model to refine an image, as shown in *Figure 6.1*. The refiner model is optional:

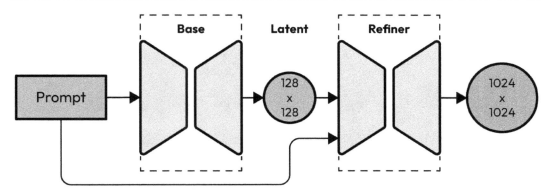

Figure 6.1: SDXL, a two-model pipeline

Figure 6.1 shows that to generate images of the best quality from the SDXL model, we will need to use the base model to generate a raw image, output as a 128x128 latent, and then use the refiner model to enhance it.

Before trying out the SDXL model, please ensure you have at least 15 GB of VRAM, otherwise, you may see a CUDA out of memory error right before the refiner model outputs the image. You can also use the optimization methods from *Chapter 5*, to build a custom pipeline to move the model out of VRAM whenever possible.

Here are the steps to load up an SDXL model:

1. Download the base model safetensors file [6]. You don't need to download all files from the model repository. At the time of writing this, the checkpoint name is sd_xl_base_1.0.safetensors.

2. Download the refiner model safetensors file [7]. We can also let the Diffusers pipeline download the safetensors file for us by providing the model ID.

3. Next, we will initialize the base and refiner models from the safetensors files:

```python
import torch
from diffusers import (
    StableDiffusionXLPipeline, StableDiffusionXLImg2ImgPipeline)

# load base model
base_model_checkpoint_path = \
    r"path/to/sd_xl_base_1.0.safetensors"
base_pipe = StableDiffusionXLPipeline.from_single_file(
    base_model_checkpoint_path,
    torch_dtype = torch.float16,
    use_safetensors = True
)
```

```
# load refiner model
refiner_model_checkpoint_path = \
    r"path/to/sd_xl_refiner_1.0.safetensors"
refiner_pipe = \
    StableDiffusionXLImg2ImgPipeline.from_single_file(
    refiner_model_checkpoint_path,
    torch_dtype = torch.float16,
    use_safetensors = True
)
```

Or, we can initialize the base and refiner models using model ID:

```
import torch
from diffusers import (
    StableDiffusionXLPipeline,
    StableDiffusionXLImg2ImgPipeline
)

# load base model
base_model_id = "stabilityai/stable-diffusion-xl-base-1.0"
base_pipe = StableDiffusionXLPipeline.from_pretrained(
    base_model_id,
    torch_dtype = torch.float16
)

# load refiner model
refiner_model_id = "stabilityai/stable-diffusion-xl-refiner-1.0"
refiner_pipe = StableDiffusionXLImg2ImgPipeline.from_pretrained(
    refiner_model_id,
    torch_dtype = torch.float16
)
```

4. Let's generate the base image in latent space (the 4x128x128 middle layer latent):

```
# move model to cuda and generate base image latent
from diffusers import EulerDiscreteScheduler

prompt = """
analog photograph of a cat in a spacesuit taken inside the
cockpit of a stealth fighter jet,
Fujifilm, Kodak Portra 400, vintage photography
```

```
"""

neg_prompt = """
paint, watermark, 3D render, illustration, drawing,worst
quality, low quality
"""

base_pipe.to("cuda")
base_pipe.scheduler = EulerDiscreteScheduler.from_config(
    base_pipe.scheduler.config)
with torch.no_grad():
    base_latents = base_pipe(
        prompt = prompt,
        negative_prompt = neg_prompt,
        output_type = "latent"
    ).images[0]

base_pipe.to("cpu")
torch.cuda.empty_cache()
```

Note that at the end of the preceding code, we moved `base_pipe` out of VRAM by using `base_pipe.to("cpu")` and `torch.cuda.empty_cache()`.

5. Load the refiner model to VRAM and use the base image in latent space to generate the final image:

```
# refine the image
refiner_pipe.to("cuda")
refiner_pipe.scheduler = EulerDiscreteScheduler.from_config(
    refiner_pipe.scheduler.config)
with torch.no_grad():
    image = refiner_pipe(
        prompt = prompt,
        negative_prompt = neg_prompt,
        image = [base_latents]
    ).images[0]

refiner_pipe.to("cpu")
torch.cuda.empty_cache()
image
```

The result will be similar to the one shown in *Figure 6.2*:

Figure 6.2: Image generated by SDXL – a cat in a spacesuit

The detail and quality are way better than the one rendered by Stable Diffusion 1.5. While this model is relatively new at the time of writing, in the near future, more mixed checkpoint models and Low-Rank Adapters (LoRAs) will be available.

Summary

This chapter mainly focused on the usage of the Stable Diffusion model. We can utilize a model from Hugging Face by using its model ID. Additionally, widely distributed open source models are available on community websites such as CIVITAI [4], where you can download numerous model resources. These model files are typically in the `.ckpt` or `safetensors` file format.

The chapter covered the distinction between these model files and using checkpoint model files directly from the `Diffusers` package. Furthermore, it offered a solution to convert standalone model checkpoint files to the Diffusers format for faster model loading.

Lastly, this chapter also covered how to load and use SDXL's two-model pipelines.

References

1. Hugging Face Load safetensors: `https://huggingface.co/docs/diffusers/using-diffusers/using_safetensors`

2. pickle — Python object serialization: `https://docs.python.org/3/library/pickle.html`

3. Stable Diffusion Safety Checker: `https://huggingface.co/CompVis/stable-diffusion-safety-checker`

4. civitai: https://www.civitai.com

5. stability.ai: https://stability.ai/

6. stable-diffusion-xl-base-1.0: https://huggingface.co/stabilityai/stable-diffusion-xl-base-1.0

7. stable-diffusion-xl-refiner-1.0: https://huggingface.co/stabilityai/stable-diffusion-xl-refiner-1.0

8. safetensors GitHub repository: https://github.com/huggingface/safetensors

9. Alec Radford et al, Learning Transferable Visual Models From Natural Language Supervision: https://arxiv.org/abs/2103.00020

10. OpenAI CLIP GitHub repository: https://github.com/openai/CLIP

11. Issues with safety checker: https://github.com/huggingface/diffusers/issues/845, https://github.com/huggingface/diffusers/issues/3422

Part 2 – Improving Diffusers with Custom Features

In Part 1, we explored the fundamental concepts and techniques behind diffusers, setting the stage for their application in various domains. Now, it's time to take our understanding to the next level by delving into advanced customization options that can significantly enhance the capabilities of these powerful models.

The chapters in this section are designed to equip you with the knowledge and skills necessary to optimize and extend your diffusers, unlocking new possibilities for creative expression and problem-solving. From refining performance and managing VRAM usage to leveraging community-driven resources and exploring innovative techniques such as textual inversion, we'll cover a range of topics that will help you push the boundaries of what's possible with diffusers.

Through the following chapters, you'll learn how to overcome limitations, tap into the collective wisdom of the community, and unlock new features that will elevate your work with diffusers. Whether you're seeking to improve efficiency, explore new artistic avenues, or simply stay at the forefront of innovation, the custom features and techniques presented in this part of the book will provide you with the tools and inspiration you need to succeed.

This part contains the following chapters:

- *Chapter 7, Optimizing Performance and VRAM Usage*
- *Chapter 8, Using Community-Shared LoRAs*
- *Chapter 9, Using Textual Inversion*
- *Chapter 10, Unlocking 77 Token Limitations and Enabling Prompt Weighting*
- *Chapter 11, Image Restore and Super-Resolution*
- *Chapter 12, Scheduled Prompt Parsing*

7

Optimizing Performance and VRAM Usage

In the previous chapters, we covered the theory behind the Stable Diffusion models, introduced the Stable Diffusion model data format, and discussed conversion and model loading. Even though the Stable Diffusion model conducts denoising in the latent space, by default, the model's data and execution still require a lot of resources and may throw a CUDA Out of memory error from time to time.

To enable fast and smooth image generation using Stable Diffusion, there are some techniques to optimize the overall process, boost the inference speed, and also reduce VRAM usage. In this chapter, we are going to cover the following optimization solutions and discuss how well these solutions work in practice:

- Using float16 or bfloat16 data type
- Enabling VAE tiling
- Enabling Xformers or using PyTorch 2.0
- Enabling sequential CPU offload
- Enabling model CPU offload
- **Token merging** (ToMe)

By using some of these solutions, you can enable your GPU with even 4 GB RAM to run a Stable Diffusion model smoothly. Please refer to *Chapter 2* for detailed software and hardware requirements for running Stable Diffusion models.

Setting the baseline

Before heading to the optimization solutions, let's take a look at the speed and VRAM usage with the default settings so that we know how much VRAM usage has been reduced or how much the speed has been improved after applying an optimization solution.

Let's use a non-cherry-picked number 1 as the generator seed to exclude the impacts from the randomly generated seed. The tests are conducted on an RTX 3090 with 24 GB VRAM running Windows 11, with another GPU for rendering all other windows and the UI so that the RTX 3090 can be dedicated to the Stable Diffusion pipelines:

```python
import torch
from diffusers import StableDiffusionPipeline

text2img_pipe = StableDiffusionPipeline.from_pretrained(
    "runwayml/stable-diffusion-v1-5"
).to("cuda:0")

# generate an image
prompt ="high resolution, a photograph of an astronaut riding a horse"
image = text2img_pipe(
    prompt = prompt,
    generator = torch.Generator("cuda:0").manual_seed(1)
).images[0]
image
```

By default, PyTorch enables **TensorFloat32 (TF32)** mode for convolutions [4] and **float32 (FP32)** mode for matrix multiplications. The preceding code generates a 512x512 image using 8.4 GB VRAM with a generation speed of 7.51iteration/second. In the following sections, we will measure the VRAM usage and the generation speed improvements after adopting an optimization solution.

Optimization solution 1 – using the float16 or bfloat16 data type

In PyTorch, floating point tensors are created in FP32 precision by default. The TF32 data format was developed for Nvidia Ampere and later CUDA devices. TF32 can achieve faster matrix multiplications and convolutions with slightly less accurate computation [5]. Both FP32 and TF32 are historic artifact settings and are required for training, but it is rare that networks need this much numerical accuracy for inference.

Instead of using the TF32 and FP32 data types, we can load and run the Stable Diffusion model weights in float16 or bfloat16 precision to save VRAM usage and improve speed. But what are the differences between float16 and bfloat16, and which one should we use?

bfloat16 and float16 are both half-precision floating-point data formats, but they have some differences:

- **Range of values**: bfloat16 has a larger positive range than float16. The maximum positive value for bfloat16 is approximately 3.39e38, while for float16 it's approximately 6.55e4. This makes bfloat16 more suitable for models that require a large dynamic range.

- **Precision**: Both bfloat16 and float16 have a 3-bit exponent and a 10-bit mantissa (fraction). However, bfloat16 uses the leading bit as a sign bit, while float16 uses it as part of the mantissa. This means that bfloat16 has a smaller relative precision than float16, especially for small numbers.

bfloat16 is typically useful for deep neural networks. It provides a good balance between range, precision, and memory usage. It's supported by many modern GPUs and can significantly reduce memory usage and increase training speed compared to using single precision (FP32).

In Stable Diffusion, we can use bfloat16 or float16 to boost the inference speed and reduce the VRAM usage at the same time. Here is some code that loads a Stable Diffusion model with bfloat16:

```
import torch
from diffusers import StableDiffusionPipeline

text2img_pipe = StableDiffusionPipeline.from_pretrained(
    "runwayml/stable-diffusion-v1-5",
    torch_dtype = torch.bfloat16 # <- load float16 version weight
).to("cuda:0")
```

We use the `text2img_pipe` pipeline object to generate an image that uses only 4.7 GB VRAM, with 19.1 denoising iterations per second.

Note that if you are using a CPU, you should not use `torch.float16` because the CPU does not have hardware support for float16.

Optimization solution 2 – enabling VAE tiling

Stable Diffusion VAE tiling is a technique that can be used to generate large images. It works by splitting an image into small tiles and then generating each tile separately. This technique allows the generation of large images without using too much VRAM.

Note that the result of tiled encoding and decoding will differ unnoticeably from the non-tiled version. Diffusers' implementation of VAE tiling uses overlap tiles to blend edges to form a much smoother output.

You can turn on VAE tiling by adding the one-line code, `text2img_pipe.enable_vae_tiling()`, before inferencing:

```
import torch
from diffusers import StableDiffusionPipeline

text2img_pipe = StableDiffusionPipeline.from_pretrained(
    "runwayml/stable-diffusion-v1-5",
    torch_dtype = torch.float16        # <- load float16 version weight
).to("cuda:0")
```

```
text2img_pipe.enable_vae_tiling()        # < Enable VAE Tiling
prompt ="high resolution, a photograph of an astronaut riding a horse"
image = text2img_pipe(
    prompt = prompt,
    generator = torch.Generator("cuda:0").manual_seed(1),
    width = 1024,
    height= 1024
).images[0]
image
```

Turning VAE tiling on or off does not seem to have much impact on the generated image. The only difference is that the VRAM usage, without VAE tiling, generates a 1024x1024 image that takes 7.6 GB VRAM. On the other hand, turning on the VAE tiling reduces the VRAM usage to 5.1 GB.

The VAE tiling happens between the image pixel space and latent space, and the overall process has a minimal impact on the denoising loop. Testing shows in the case of generating fewer than 4 images, there is an unnoticeable impact on the performance, which can reduce VRAM usage by 20% to 30%. It would be a good idea to always turn it on.

Optimization solution 3 – enabling Xformers or using PyTorch 2.0

When we provide a text or prompt to generate an image, the encoded text embedding will be fed to the Transformer multi-header attention component of the diffusion UNet.

Inside the Transformer block, the self-attention and cross-attention headers will try to compute the attention score (via the QKV operation). This is computation-heavy and will also use a lot of memory.

The open source Xformers [2] package from Meta Research is built to optimize the process. In short, the main differences between Xformers and standard Transformers are as follows:

- **Hierarchical attention mechanism**: Xformers use a hierarchical attention mechanism, which consists of two layers of attention: a coarse layer and a fine layer. The coarse layer attends to the input sequence at a high level, while the fine layer attends to the input sequence at a low level. This allows Xformers to learn long-range dependencies in the input sequence while also being able to focus on local details.

- **Reduced number of heads**: Xformers use a smaller number of heads than standard Transformers. A head is a unit of computation in the attention mechanism. Xformers use 4 heads, while standard Transformers use 12 heads. This reduction in the number of heads allows Xformers to reduce the memory requirements while still maintaining performance.

Enabling Xformers for Stable Diffusion using the `Diffusers` package is quite simple. Add one line of code, as shown in the following snippet:

```
import torch
from diffusers import StableDiffusionPipeline

text2img_pipe = StableDiffusionPipeline.from_pretrained(
    "runwayml/stable-diffusion-v1-5",
    torch_dtype = torch.float16        # <- load float16 version weight
).to("cuda:0")

text2img_pipe.enable_xformers_memory_efficient_attention()  # < Enable
# xformers
prompt ="high resolution, a photograph of an astronaut riding a horse"
image = text2img_pipe(
    prompt = prompt,
    generator = torch.Generator("cuda:0").manual_seed(1)
).images[0]
image
```

If you are using PyTorch 2.0+, you may not notice the performance boost or VRAM usage drop. That is because PyTorch 2.0 includes a natively built attention optimization feature similar to the Xformers implementation. If a historical PyTorch before version 2.0 is being used, enabling Xformers will noticeably boost the inference speed and reduce VRAM usage.

Optimization solution 4 – enabling sequential CPU offload

As we discussed in *Chapter 5*, one pipeline includes several sub-models:

- Text embedding model used to encode text to embeddings
- Image latent encoder/decoder used to encode the input guidance image and decode latent space to pixel images
- The UNet will loop the inference denoising steps
- The safety checker model checks the safety of the generated content

The idea of sequential CPU offload is offloading idle submodels to CPU RAM when it finishes its task and is idle.

Here is an example of how it works step by step:

1. Load the CLIP text model to the GPU VRAM and encode the input prompt to embeddings.
2. Offload the CLIP text model to CPU RAM.

3. Load the VAE model (the image to latent space encoder and decoder) to the GPU VRAM and encode the start image if the current task is an image-to-image pipeline.

4. Offload the VAE to the CPU RAM.

5. Load UNet to loop through the denoising steps (load and offload unused sub-modules weights data too).

6. Offload UNet to the CPU RAM.

7. Load the VAE model from CPU RAM to GPU VRAM to perform latent space to image decoding.

In the preceding steps, we can see that through all the processes, only one sub-model will stay in the VRAM, which can effectively reduce the usage of VRAM. However, the loading and offloading will significantly reduce the inference speed.

Enabling sequential CPU offload is as simple as one line of code, as shown in the following snippet:

```
import torch
from diffusers import StableDiffusionPipeline

text2img_pipe = StableDiffusionPipeline.from_pretrained(
    "runwayml/stable-diffusion-v1-5",
    torch_dtype = torch.float16
).to("cuda:0")

# generate an image
text2img_pipe.enable_sequential_cpu_offload() # <- Enable sequential
# CPU offload
prompt ="high resolution, a photograph of an astronaut riding a horse"
image = text2img_pipe(
    prompt = prompt,
    generator = torch.Generator("cuda:0").manual_seed(1)
).images[0]
image
```

Imagine the possibility of creating a tailored pipeline that efficiently utilizes the VRAM for denoising with UNet. By strategically shifting the text encoder/decoder, VAE models, and safety checker models to the CPU during idle periods, while keeping the UNet model in the VRAM, significant speed enhancements are achievable. The feasibility of this approach is evident in the custom implementation provided in the code that comes with the book, which remarkably reduces VRAM usage to as low as 3.2 GB (even for generating a 512x512 image) while maintaining a comparable processing speed, with no noticeable degradation in performance!

The custom pipeline code provided in this chapter did almost the same thing as enable_sequential_ cpu_offload(). The only difference is keeping the UNet in VRAM until the end of denoising. That is why the inference speed remains fast.

With proper model load and offload management, we can reduce the VRAM usage from 4.7 GB to 3.2 GB while maintaining inference speeds that are indistinguishable from those achieved without model offloading.

Optimization solution 5 – enabling model CPU offload

Full model offloading moves the whole model data to and off GPU instead of moving weights only. If this is not enabled, all model data will stay in GPU before and after forward inference; clearing the CUDA cache won't free up VRAM either. This could lead to a CUDA Out of memory error if you are loading up other models, say, an upscale model to further process the image. The model-to-CPU offload method can mitigate the CUDA Out of memory problem.

Based on the idea behind this method, an additional one to two seconds will be spent on moving the model between CPU RAM and GPU VRAM.

To enable this method, remove pipe.to("cuda") and add pipe.enable_model_cpu_ offload():

```
import torch
from diffusers import StableDiffusionPipeline

text2img_pipe = StableDiffusionPipeline.from_pretrained(
    "runwayml/stable-diffusion-v1-5",
    torch_dtype = torch.float16
)                 # .to("cuda") is removed here

# generate an image
text2img_pipe.enable_model_cpu_offload()     # <- enable model offload
prompt ="high resolution, a photograph of an astronaut riding a horse"
image = text2img_pipe(
    prompt = prompt,
    generator = torch.Generator("cuda:0").manual_seed(1)
).images[0]
image
```

When offloading the model, the GPU hosts a single primary pipeline component, usually the text encoder, UNet, or VAE, while the remaining components are idle on the CPU memory. Components such as UNet, which undergo multiple iterations, remain on the GPU until their utilization is no longer required.

The model CPU offload method can reduce the VRAM usage to 3.6 GB and keep a relatively good inference speed. If you give the preceding code a test run, you will find the inference speed is relatively slow at the beginning and gradually speeds up to its normal iteration speed.

At the end of image generation, we can use the following code to manually move the model weights data out of VRAM to CPU RAM:

```
pipe.to("cpu")
torch.cuda.empty_cache()
```

After executing the preceding code, you will find your GPU VRAM usage level has significantly reduced.

Next, let's take a look at Token Merging.

Optimization solution 6 – Token Merging (ToMe)

Token Merging (**ToMe**) was first posited by Daniel et al [3]. It is a technique that can be used to speed up the inference time of Stable Diffusion models. ToMe works by merging redundant tokens in the model, which means that the model has less work to do compared with non-merging models. This can lead to noticeable speed improvements without sacrificing image quality.

ToMe works by first identifying redundant tokens in the model. This is done by looking at the similarity between tokens. If two tokens are very similar, then they are probably redundant. Once redundant tokens have been identified, they are merged. This is done by averaging the values of the two tokens.

For example, if a model has 100 tokens and 50 of those tokens are redundant, then merging the redundant tokens can reduce the number of tokens that the model has to process by 50%.

ToMe can be used with any Stable Diffusion model. It does not require any additional training. To use ToMe, we need to first install the following package from its original inventor:

```
pip install tomesd
```

Then, import the ToMe package to enable it:

```
import torch
from diffusers import StableDiffusionPipeline
import tomesd

text2img_pipe = StableDiffusionPipeline.from_pretrained(
    "runwayml/stable-diffusion-v1-5",
    torch_dtype = torch.float16
).to("cuda:0")

tomesd.apply_patch(text2img_pipe, ratio=0.5)
# generate an image
prompt ="high resolution, a photograph of an astronaut riding a horse"
image = text2img_pipe(
    prompt = prompt,
```

```
    generator = torch.Generator("cuda:0").manual_seed(1)
).images[0]
image
```

The performance improvement is dependent on how many redundant tokens are found. In the preceding code, the ToMe package improves the iteration speed from around 19 iterations per second to 20 iterations per second.

It's worth noting that the ToMe package may produce a slightly altered image output, although this difference has no discernible impact on image quality. This is because ToMe merges tokens, which can influence the conditional embeddings.

Summary

In this chapter, we have introduced six techniques to enhance the performance of Stable Diffusion and minimize VRAM usage. The amount of VRAM is often the most significant hurdle in running a Stable Diffusion model, with CUDA Out of memory being a common issue. The techniques we have discussed can drastically reduce VRAM usage while maintaining the same inference speed.

Enabling the float16 data type can halve VRAM usage and nearly double the inference speed. VAE tiling allows the generation of large images without excessive VRAM usage. Xformers can further decrease VRAM usage and increase inference speed by implementing an intelligent two-layer attention mechanism. PyTorch 2.0 provides native features such as Xformers and automatically enables them.

Sequential CPU offload can significantly reduce VRAM usage by offloading a sub-model and its sub-modules to CPU RAM, albeit at the cost of slower inference speed. However, we can use the same concept to implement our sequential offload mechanism to save VRAM usage while keeping the inference speed nearly the same. Model CPU offload can offload the entire model to the CPU, freeing up VRAM for other tasks, and only reloading the models back to VRAM when necessary. **Token Merging**, or **ToMe**, reduces redundant tokens and boosts inference speed.

By applying these solutions, you could potentially run a pipeline that outperforms any other models in the world. The AI landscape is constantly evolving, and by the time you read this, new solutions may have emerged. However, understanding the internal workings allows us to tune and optimize the image generation process according to your needs.

In the next chapter, we are going to explore of the most exciting topics, community-shared LoRAs.

References

1. Hugging Face, memory, and speed: `https://huggingface.co/docs/diffusers/optimization/fp16`

2. facebookresearch, xformers: `https://github.com/facebookresearch/xformers`

3. Daniel Bolya, Judy Hoffman; Token Merging for Fast Stable Diffusion: `https://arxiv.org/abs/2303.17604`

4. What Every User Should Know About Mixed Precision Training in PyTorch: `https://pytorch.org/blog/what-every-user-should-know-about-mixed-precision-training-in-pytorch/#picking-the-right-approach`

5. Accelerating AI Training with NVIDIA TF32 Tensor Cores: `https://developer.nvidia.com/blog/accelerating-ai-training-with-tf32-tensor-cores/`

8

Using Community-Shared LoRAs

To meet specific needs and generate higher fidelity images, we may need to fine-tune a pre-trained Stable Diffusion model, but the fine-tuning process is extremely slow without powerful GPUs. Even if you have all the hardware or resources on hand, the fine-tuned model is large, usually the same size as the original model file.

Fortunately, researchers from the Large Language Model (LLM) neighbor community developed an efficient fine-tuning method, **Low-Rank Adaptation (LoRA** — "Low" is why the "o" is in lowercase) [1]. With LoRA, the original checkpoint is frozen without any modification, and the tuned weight changes are stored in an independent file, which we usually call the LoRA file. Additionally, there are countless community-shared LoRAs on sites such as CIVITAI [4] and HuggingFace.

In this chapter, we are going to delve into the theory of LoRA, and then introduce the Python way to load up LoRA into a Stable Diffusion model. We will also dissect a LoRA model to understand the LoRA model structure internally and create a custom function to load up a Stable Diffusion V1.5 LoRA.

The following topics will be covered in this chapter:

- How does LoRA work?
- Using LoRA with Diffusers
- Applying LoRA weight during loading
- Diving into LoRA
- Making a function to load LoRA
- Why LoRA works

By the end of this chapter, we will be able to use any community LoRA programmatically and also understand how and why LoRA works in Stable Diffusion.

Technical requirements

If you have `Diffusers` package running in your computer, you should be able to execute all code in this chapter as well as the code used to load LoRA with Diffusers.

Diffusers use **PEFT (Parameter-Efficient Fine-Tuning)** [10] to manage the LoRA loading and offloading. PEFT is a library developed by Hugging Face that provides parameter-efficient ways to adapt large pre-trained models for specific downstream applications. The key idea behind PEFT is to fine-tune only a small fraction of a model's parameters instead of fine-tuning all of them, resulting in significant savings in terms of computation and memory usage. This makes it possible to fine-tune very large models even on consumer hardware with limited resources. Turn to *Chapter 21* for more about LoRA.

We will need to install the PEFT package to enable Diffusers' PEFT LoRA loading:

```
pip install PEFT
```

You can also refer to *Chapter 2*, if you encounter other execution errors from the code.

How does LoRA work?

LoRA is a technique for quickly fine-tuning diffusion models, first introduced by Microsoft researchers in a paper by Edward J. Hu et al [1]. It works by creating a small, low-rank model that is adapted for a specific concept. This small model can be merged with the main checkpoint model to generate images similar to the ones used to train LoRA.

Let's use W to denote the original UNet attention weights (Q,K,V), ΔW to denote the fine-tuned weights from LoRA, and W' as the merged weights. The process of adding LoRA to a model can be expressed like this:

$$W' = W + \Delta W$$

If we want to control the scale of LoRA weights, we denote the scale as α. Adding LoRA to a model can be expressed like this now:

$$W' = W + \alpha \Delta W$$

The range of α can be from 0 to 1.0 [2]. It should be fine if we set α slightly larger than 1.0. The reason why LoRA is so small is that ΔW can be represented by two small matrices A and B, such that:

$$\Delta W = AB^T$$

Where $A \in \mathbb{R}^{n \times d}$ is an $n \times d$ matrix, and $B \in \mathbb{R}^{m \times d}$ is an $m \times d$ matrix. The transpose of B denoted as B^T is a $d \times m$ matrix.

For example, if ΔW is a 6×8 matrix, there are a total of 48 weight numbers. Now, in the LoRA file, the 6×8 matrix can be represented by two matrices – one 6×2 matrix, 12 numbers in total, and another 2×8 matrix, making it 16 numbers in total.

The total number of weights is reduced from 48 to 28. This is why the LoRA file can be so small compared to the checkpoint model.

Using LoRA with Diffusers

With the contributions from the open source community, loading LoRA with Python has never been easier. In this section, we are going to cover the solutions to load a LoRA model with Diffusers.

In the following steps, we will first load the base Stable Diffusion V1.5, generate an image without LoRA, and then load a LoRA model called MoXinV1 into the base model. We will clearly see the difference with and without the LoRA model:

1. **Prepare a Stable Diffusion pipeline**: The following code will load up the Stable Diffusion pipeline and move the pipeline instance to VRAM:

    ```
    import torch
    from diffusers import StableDiffusionPipeline

    pipeline = StableDiffusionPipeline.from_pretrained(
        "runwayml/stable-diffusion-v1-5",
        torch_dtype = torch.float16
    ).to("cuda:0")
    ```

2. **Generate an image without LoRA**: Now, let's generate an image without LoRA loaded. Here, I am going to use the Stable Diffusion default v1.5 model to generate "a branch of flower" in a "traditional Chinese ink painting" style:

    ```
    prompt = """
    shukezouma, shuimobysim, a branch of flower, traditional chinese
    ink painting
    """
    image = pipeline(
        prompt = prompt,
        generator = torch.Generator("cuda:0").manual_seed(1)
    ).images[0]
    display(image)
    ```

    The preceding code uses a non-cherry-picked generator with default seed:1. The result is shown in *Figure 8.1*:

Figure 8.1: A branch of flower without LoRA

To be honest, the preceding image isn't that good, and the "flower" is more like black ink dots.

3. **Generate an image with LoRA with default settings**: Next, let's load up the LoRA model to the pipeline and see what the MoXin LoRA can do to help image generation. Loading LoRA with default settings is just one line of code:

```
# load LoRA to the pipeline
pipeline.load_lora_weights(
    "andrewzhu/MoXinV1",
    weight_name  = "MoXinV1.safetensors",
    adapter_name = "MoXinV1"
)
```

Diffusers downloads the LoRA model file automatically if the model does not exist in your model cache.

Now, run the inference again with the following code (the same code used in *step 2*):

```
image = pipeline(
    prompt = prompt,
    generator = torch.Generator("cuda:0").manual_seed(1)
).images[0]
display(image)
```

We will have a new image with a better "flower" in the ink-painting style, as shown in *Figure 8.2*:

Figure 8.2: A branch of flower with LoRA using the default settings

This time, the "flower" is more like a flower and, overall, better than the one without applying LoRA. However, the code in this section loads LoRA without a "weight" applied to it. In the next section, we will load a LoRA model with an arbitrary weight (or α).

Applying a LoRA weight during loading

In the *How does LoRA work?* section, we mentioned the α value used to define the portion of LoRA weight added to the main model. We can easily achieve this using Diffusers with PEFT [10].

What is PEFT? PEFT is a library developed by Hugging Face to efficiently adapt pre-trained models, such as **Large Language Models (LLMs)** and Stable Diffusion models, to new tasks without needing to fine-tune the whole model. PEFT is a broader concept representing a collection of methods aimed at efficiently fine-tuning LLMs. LoRA, conversely, is a specific technique that falls under the umbrella of PEFT.

Before the integration of PEFT, loading and managing LoRAs in Diffusers required a lot of custom code and hacking. To make it easier to manage multiple LoRAs with weight loading and offloading, Diffusers uses the PEFT library to help manage different adapters for inference. In PEFT, the fine-tuned parameters are called adapters, which is why you will see some parameters are named `adapters`. LoRA is one of the main adapter techniques; you can take LoRA and adapter as referring to the same thing through this chapter.

Loading a LoRA model with weight is simple, as shown in the following code:

```
pipeline.set_adapters(
    ["MoXinV1"],
    adapter_weights=[0.5]
)
image = pipeline(
    prompt = prompt,
    generator = torch.Generator("cuda:0").manual_seed(1)
).images[0]
display(image)
```

In the preceding code, we gave the LoRA weight as 0.5 to replace the default 1.0. Now, you will see the generated image, as shown in *Figure 8.3*:

Figure 8.3: A branch of flower with LoRA by applying a 0.5 LoRA weight

From *Figure 8.3*, we can observe the difference after applying the 0.5 weight to the LoRA model.

The PEFT-integrated Diffusers can also load another LoRA by reusing the same code we used to load the first LoRA model:

```
# load another LoRA to the pipeline
pipeline.load_lora_weights(
    "andrewzhu/civitai-light-shadow-lora",
```

```
        weight_name     = "light_and_shadow.safetensors",
        adapter_name    = "light_and_shadow"
)
```

Then, add the weight for the second LoRA model by calling the `set_adapters` function:

```
pipeline.set_adapters(
    ["MoXinV1", "light_and_shadow"],
    adapter_weights=[0.5,1.0]
)
prompt = """
shukezouma, shuimobysim ,a branch of flower, traditional chinese ink
painting,STRRY LIGHT,COLORFUL
"""
image = pipeline(
    prompt = prompt,
    generator = torch.Generator("cuda:0").manual_seed(1)
).images[0]
display(image)
```

We will get a new image with style added from the second LoRA, as shown in *Figure 8.4*:

Figure 8.4: A branch of flower with two LoRA models

We can also use the same code to load LoRA for Stable Diffusion XL pipelines.

With PEFT, we don't need to restart the pipeline to disable LoRA; we can disable all LoRAs with simply one line of code:

```
pipeline.disable_lora()
```

Note that the implementation of LoRA loading is somewhat different compared with other tools, such as A1111 Stable Diffusion WebUI. Using the same prompt, the same settings, and the same LoRA weight, you may get a different result.

Don't worry – in the next section, we are going to dive into the LoRA model internally and implement a solution to use LoRA that outputs the same result, with tools such as A1111 Stable Diffusion WebUI.

Diving into the internal structure of LoRA

Understanding how LoRA works internally will help us to implement our own LoRA-related features based on specific needs. In this section, we are going to dive into the internals of LoRA's structure and its weights schema, and then manually load the LoRA model into the Stable Diffusion model step by step.

As we discussed at the beginning of the chapter, applying LoRA is as simple as the following:

$$W' = W + \alpha \Delta W$$

And ΔW can be broken down into A and B:

$$\Delta W = A B^T$$

So, the overall idea of merging LoRA weights to the checkpoint model works like this:

1. Find the A and B weight matrix from the LoRA file.
2. Match the LoRA module layer name to the checkpoint module layer name so that we know which matrix to merge.
3. Produce $\Delta W = A B^T$.
4. Update the checkpoint model weights.

If you have prior experience training a LoRA model, you might be aware that a hyperparameter, `alpha`, can be set to a value greater than 1, such as 4. This is often done in conjunction with setting another parameter, `rank`, to 4 as well. However, α used in this context is typically less than 1. The actual value of α is generally computed using the following formula:

$$\alpha = \frac{alpha}{rank}$$

During the training phase, setting both `alpha` and `rank` to 4 will yield an α value of 1. This concept may seem confusing if not properly understood.

Next, let's explore the internals of a LoRA model step by step.

Finding the *A* and *B* weight matrix from the LoRA file

Before start exploring the internals of a LoRA structure, you will need to download a LoRA file. You can download the MoXinV1.safetensors from the following URL: https://huggingface. co/andrewzhu/MoXinV1/resolve/main/MoXinV1.safetensors.

After setting up the LoRA file in the .safetensors format, load it using the following code:

```
# load lora file
from safetensors.torch import load_file
lora_path = "MoXinV1.safetensors"
state_dict = load_file(lora_path)
for key in state_dict:
    print(key)
```

When LoRA weights are applied to the text encoder, the key names start with lora_te_:

```
. . .
lora_te_text_model_encoder_layers_7_mlp_fc1.alpha
lora_te_text_model_encoder_layers_7_mlp_fc1.lora_down.weight
lora_te_text_model_encoder_layers_7_mlp_fc1.lora_up.weight
. . .
```

When LoRA weights are applied to UNet, key names start with lora_unet_:

```
. . .
lora_unet_down_blocks_0_attentions_1_proj_in.alpha
lora_unet_down_blocks_0_attentions_1_proj_in.lora_down.weight
lora_unet_down_blocks_0_attentions_1_proj_in.lora_up.weight
. . .
```

The output is of the string type. Here are the meanings of the terms that appeared in the output LoRA weight keys:

- The lora_te_ prefix says that the weights are applied to the text encoder; lora_unet_ says that the weights aim at updating the Stable Diffusion unet module layers.
- down_blocks_0_attentions_1_proj_in is the layer name, which should exist in the checkpoint model unet modules too.
- .alpha is the trained weight set to denote how much of the LoRA weight will be applied to the main checkpoint model. It holds a float value that is denoted as α in $W' = W + \alpha \Delta W$. Since the value will be replaced by user input, we can skip this value.
- lora_down.weight denotes the value of this layer that represents *A*.

- `lora_up.weight` denotes the value of this layer that represents *B*.

- Note that `down` in `down_blocks` denotes the downside (the left side of UNet) of the unet model.

The following Python code will get the LoRA layer info and also have the model object handler:

```
# find the layer name
LORA_PREFIX_UNET = 'lora_unet'
LORA_PREFIX_TEXT_ENCODER = 'lora_te'
for key in state_dict:
    if 'text' in key:
        layer_infos = key.split('.')[0].split(
            LORA_PREFIX_TEXT_ENCODER+'_')[-1].split('_')
        curr_layer = pipeline.text_encoder
    else:
        layer_infos = key.split('.')[0].split(
            LORA_PREFIX_UNET+'_')[-1].split('_')
        curr_layer = pipeline.unet
```

`key` holds the LoRA module layer name, and `layer_infos` holds the checkpoint model layer name extracted from the LoRA layers. The reason we do this is that not all layers from the checkpoint model have LoRA weights to adjust, which is why we need to get the list of layers that will be updated.

Finding the corresponding checkpoint model layer name

Print out the structure of the checkpoint model `unet` structure:

```
unet = pipeline.unet
modules = unet.named_modules()
for child_name, child_module in modules:
    print("child_module:",child_module)
```

We can see that the module is stored in a tree structure like this:

```
...
(down_blocks): ModuleList(
    (0): CrossAttnDownBlock2D(
        (attentions): ModuleList(
        (0-1): 2 x Transformer2DModel(
            (norm): GroupNorm(32, 320, eps=1e-06, affine=True)
            (proj_in): Conv2d(320, 320, kernel_size=(1, 1), stride=(1,
1))
            (transformer_blocks): ModuleList(
            (0): BasicTransformerBlock(
                (norm1): LayerNorm((320,), eps=1e-05, elementwise_
```

```
affine=True)
                (attn1): Attention(
                (to_q): Linear(in_features=320, out_features=320,
bias=False)
                (to_k): Linear(in_features=320, out_features=320,
bias=False)
                (to_v): Linear(in_features=320, out_features=320,
bias=False)
                (to_out): ModuleList(
                    (0): Linear(in_features=320, out_features=320,
bias=True)
                    (1): Dropout(p=0.0, inplace=False)
                )
...
```

Each line is composed of a module name (`down_blocks`), and the module content can be `ModuleList` or a specific neural network layer, `Conv2d`. These are the components of the UNet. For now, applying LoRA to a specific UNet module isn't required. However, it's important to understand the UNet's internal structure:

```
# find the layer name
for key in state_dict:
    # find the LoRA layer name (the same code shown above)
    for key in state_dict:
    if 'text' in key:
        layer_infos = key.split('.')[0].split(
            "lora_unet_")[-1].split('_')
        curr_layer = pipeline.text_encoder
    else:
        layer_infos = key.split('.')[0].split(
            "lora_te_")[-1].split('_')
        curr_layer = pipeline.unet

    # loop through the layers to find the target layer
    temp_name = layer_infos.pop(0)
    while len(layer_infos) > -1:
        try:
            curr_layer = curr_layer.__getattr__(temp_name)
            # no exception means the layer is found
            if len(layer_infos) > 0:
                temp_name = layer_infos.pop(0)
            # all names are pop out, break out from the loop
            elif len(layer_infos) == 0:
                break
        except Exception:
```

```
            # no such layer exist, pop next name and try again
            if len(temp_name) > 0:
                temp_name += '_'+layer_infos.pop(0)
            else:
                # temp_name is empty
                temp_name = layer_infos.pop(0)
```

The loop-through part is a bit tricky. When looking back to the checkpoint model structure, which is layered as a tree, we can't simply use a `for` loop to loop through the list. Instead, we need to use a `while` loop to navigate every leaf of the tree. The overall process is as follows:

1.   `layer_infos.pop(0)` will return the first name of the list in the `string` type and remove it from the list such as up from the `layer_infos` list – `['up', 'blocks', '3', 'attentions', '2', 'transformer', 'blocks', '0', 'ff', 'net', '2']`

2.   Use `curr_layer.__getattr__(temp_name)` to check whether the layer exists or not. If it does not exist, an exception will be thrown, and the program will move to the `exception` section to continue outputting names from the `layer_infos` list and check again.

3.   If the layer is found but some names are still left in the `layer_infos` list, they will keep on popping out.

4.   The names will continue to pop out until no exception is thrown out and we meet the `len(layer_infos) == 0` condition, which means that the layer is fully matched.

At this point, the `curr_layer` object points to the checkpoint model weight data and can be referenced in the next step.

Updating the checkpoint model weights

For easier key value referencing, let's make a `pair_keys = []` list, in which `pair_keys[0]` returns the *A* matrix and `pair_keys[1]` returns the *B* matrix:

```
# ensure the sequence of lora_up(A) then lora_down(B)
pair_keys = []
if 'lora_down' in key:
    pair_keys.append(key.replace('lora_down', 'lora_up'))
    pair_keys.append(key)
else:
    pair_keys.append(key)
    pair_keys.append(key.replace('lora_up', 'lora_down'))
```

Then, we update the weights:

```
alpha = 0.5
# update weight
```

```
if len(state_dict[pair_keys[0]].shape) == 4:
    # squeeze(3) and squeeze(2) remove dimensions of size 1
    #from the tensor to make the tensor more compact
    weight_up = state_dict[pair_keys[0]].squeeze(3).squeeze(2).\
        to(torch.float32)
    weight_down = state_dict[pair_keys[1]].squeeze(3).squeeze(2).\
        to(torch.float32)
    curr_layer.weight.data += alpha * torch.mm(weight_up,
        weight_down).unsqueeze(2).unsqueeze(3)
else:
    weight_up = state_dict[pair_keys[0]].to(torch.float32)
    weight_down = state_dict[pair_keys[1]].to(torch.float32)
    curr_layer.weight.data += alpha * torch.mm(weight_up, weight_down)
```

The `alpha * torch.mm(weight_up, weight_down)` code is the core code used to implement $\alpha A\,B^T$.

And that's it! Now, the pipeline's text encoder and `unet` model weights are updated by LoRA. Next, let's put all the parts together to create a full-featured function that can load a LoRA model into the Stable Diffusion pipeline.

Making a function to load LoRA

Let's add one more list to store keys that have been visited and put all the preceding code together into a function named `load_lora`:

```
def load_lora(
    pipeline,
    lora_path,
    lora_weight = 0.5,
    device = 'cpu'
):
    state_dict = load_file(lora_path, device=device)
    LORA_PREFIX_UNET = 'lora_unet'
    LORA_PREFIX_TEXT_ENCODER = 'lora_te'

    alpha = lora_weight
    visited = []

    # directly update weight in diffusers model
    for key in state_dict:
        # as we have set the alpha beforehand, so just skip
        if '.alpha' in key or key in visited:
            continue
```

```python
if 'text' in key:
    layer_infos = key.split('.')[0].split(
        LORA_PREFIX_TEXT_ENCODER+'_')[-1].split('_')
    curr_layer = pipeline.text_encoder
else:
    layer_infos = key.split('.')[0].split(
        LORA_PREFIX_UNET+'_')[-1].split('_')
    curr_layer = pipeline.unet

# find the target layer
# loop through the layers to find the target layer
temp_name = layer_infos.pop(0)
while len(layer_infos) > -1:
    try:
        curr_layer = curr_layer.__getattr__(temp_name)
        # no exception means the layer is found
        if len(layer_infos) > 0:
            temp_name = layer_infos.pop(0)
        # layer found but length is 0,
        # break the loop and curr_layer keep point to the
        # current layer
        elif len(layer_infos) == 0:
            break
    except Exception:
        # no such layer exist, pop next name and try again
        if len(temp_name) > 0:
            temp_name += '_'+layer_infos.pop(0)
        else:
            # temp_name is empty
            temp_name = layer_infos.pop(0)

# org_forward(x) + lora_up(lora_down(x)) * multiplier
# ensure the sequence of lora_up(A) then lora_down(B)
pair_keys = []
if 'lora_down' in key:
    pair_keys.append(key.replace('lora_down', 'lora_up'))
    pair_keys.append(key)
else:
    pair_keys.append(key)
    pair_keys.append(key.replace('lora_up', 'lora_down'))

# update weight
```

```
        if len(state_dict[pair_keys[0]].shape) == 4:
            # squeeze(3) and squeeze(2) remove dimensions of size 1
            # from the tensor to make the tensor more compact
            weight_up = state_dict[pair_keys[0]].squeeze(3).\
                squeeze(2).to(torch.float32)
            weight_down = state_dict[pair_keys[1]].squeeze(3).\
                squeeze(2).to(torch.float32)
            curr_layer.weight.data += alpha * torch.mm(weight_up,
                weight_down).unsqueeze(2).unsqueeze(3)
        else:
            weight_up = state_dict[pair_keys[0]].to(torch.float32)
            weight_down = state_dict[pair_keys[1]].to(torch.float32)
            curr_layer.weight.data += alpha * torch.mm(weight_up,
                weight_down)

        # update visited list, ensure no duplicated weight is
        # processed.
        for item in pair_keys:
            visited.append(item)
```

To use the function is easy; simply provide the `pipeline` object, the LoRA path, `lora_path`, and the LoRA weight number, `lora_weight`, like this:

```
pipeline = StableDiffusionPipeline.from_pretrained(
    "runwayml/stable-diffusion-v1-5",
    torch_dtype = torch.bfloat16
).to("cuda:0")

lora_path = r"MoXinV1.safetensors"
load_lora(
    pipeline = pipeline,
    lora_path = lora_path,
    lora_weight = 0.5,
    device = "cuda:0"
)
```

Now, let's try it out:

```
prompt = """
shukezouma, shuimobysim ,a branch of flower, traditional chinese ink
painting
"""
image = pipeline(
    prompt = prompt,
    generator = torch.Generator("cuda:0").manual_seed(1)
```

```
).images[0]
display(image)
```

It works, and works well; see the result shown in *Figure 8.5*:

Figure 8.5: A branch of flower using the custom LoRA loader

You might be wondering, "Why does a small LoRA file possess such formidable capabilities?" Let's delve deeper into the reasons why a LoRA model is effective.

Why LoRA works

The paper *Intrinsic Dimensionality Explains the Effectiveness of Language Model Fine-Tuning* [8] by Armen et al. found that the pre-trained representations' intrinsic dimension is way lower than expected, stated by them as follows:

> *"We empirically show that common NLP tasks within the context of pre-trained representations have an intrinsic dimension several orders of magnitudes less than the full parameterization."*

The intrinsic dimension of a matrix is a concept used to determine the effective number of dimensions required to represent the important information contained within that matrix.

Let's suppose we have a matrix, M, with five rows and three columns, like this:

```
M =  1   2   3
     4   5   6
     7   8   9
    10  11  12
    13  14  15
```

Each row of this matrix represents a data point or a vector with three values. We can think of these vectors as points in a three-dimensional space. However, if we visualize these points, we might find that they lie approximately on a two-dimensional plane, rather than occupying the full three-dimensional space.

In this case, the intrinsic dimension of the matrix, M, would be 2, indicating that the essential structure of the data can be captured effectively using two dimensions. The third dimension doesn't provide much additional information.

A low intrinsic dimension matrix can be represented by two low-rank matrices because the data in the matrix can be compressed into a few key features. These features can then be represented by two smaller matrices, each of which has a rank that is equal to the intrinsic dimension of the original matrix.

The paper *LoRA: Low-Rank Adaptation of Large Language Models* [1] by Edward J. Hu et al goes a step further, introducing the concept of LoRA to leverage the low intrinsic dimension nature, boosting the fine-tuning process by breaking down the delta weights to two low-rank parts, $\Delta W = A B^T$.

The effectiveness of LoRA was soon discovered to extend beyond LLM models, also yielding good results with diffusion models. Simo Ryu published the LoRA [2] code and was the first one to try out LoRA training for Stable Diffusion. That was in July 2023 and there are now more than 40,000 LoRA models shared at `https://www.civitai.com`.

Summary

In this chapter, we discussed how to enhance the Stable Diffusion model using LoRA, understood what LoRA is, and why it is good for fine-tuning and inference.

Then, we began loading LoRA using the experimental functions from the `Diffusers` package and provided LoRA weights through a custom implementation. We used simple code to quickly understand what LoRA can bring to the table.

Then, we dived into the internal structure of a LoRA model, walked through the detailed steps to extract LoRA weights, and understood how to merge those weights into the checkpoint model.

Further, we implemented a function in Python that can load a LoRA safetensors file and perform weight merges.

Finally, we briefly explored why LoRA works, based on the most recent papers from researchers.

In the next chapter, we are going to explore another powerful technique – textual inversion – to teach a model new "words," and then use the pre-trained "words" to add new concepts to the generated images.

References

1. Edward J. et al, LoRA: Low-Rank Adaptation of Large Language Models: `https://arxiv.org/abs/2106.09685`

2. Simo Ryu (cloneofsimo), `lora`: `https://github.com/cloneofsimo/lora`

3. `kohya_lora_loader`: `https://gist.github.com/takuma104/e38d683d72b1e448b8d9b3835f7cfa44`

4. CIVITAI: `https://www.civitai.com`

5. Rinon Gal et al, An Image is Worth One Word: Personalizing Text-to-Image Generation using Textual Inversion: `https://textual-inversion.github.io/`

6. Diffusers' `lora_state_dict` function: `https://github.com/huggingface/diffusers/blob/main/src/diffusers/models/modeling_utils.py`

7. Andrew Zhu, Improving Diffusers Package for High-Quality Image Generation: `https://towardsdatascience.com/improving-diffusers-package-for-high-quality-image-generation-a50fff04bdd4`

8. Armen et al, Intrinsic Dimensionality Explains the Effectiveness of Language Model Fine-Tuning: `https://arxiv.org/abs/2012.13255`

9. Hugging Face, LoRA: `https://huggingface.co/docs/diffusers/training/lora`

10. Hugging Face, PEFT: `https://huggingface.co/docs/peft/en/index`

9
Using Textual Inversion

Textual inversion (**TI**) is another way to provide additional capabilities to a pretrained model. Unlike **Low-Rank Adaptation** (**LoRA**), discussed in *Chapter 8*, which is a fine-tuning technique applied to the text encoder and the UNet attention weights, TI is a technique to add new **embedding** space based on the trained data.

In the context of Stable Diffusion, **text embedding** refers to the representation of text data as numerical vectors in a high dimensional space, allowing for manipulation and processing by machine learning algorithms. Specifically, in the case of Stable Diffusion, text embeddings are typically created using the **Contrastive Language-Image Pretraining** (**CLIP**) [6] model.

To train a TI model, you only need a minimal set of three to five images, resulting in a compact pt or bin file, typically just a few kilobytes in size. This makes TI a highly efficient method for incorporating new elements, concepts, or styles into your pretrained checkpoint model while maintaining exceptional portability.

In this chapter, we will first start using TI with the TI loader from the diffusers package, then delve into the core of TI to uncover how it works internally, and finally build a custom TI loader to have the TI weight applied to the image generation.

Here are the topics we are going to cover:

- Diffusers inference using TI
- How TI works
- Build a custom TI loader

By the end of this chapter, you will be able to start using any type of TI shared by the community and also build your application to load TI.

Let's start leveraging the power of Stable Diffusion TI.

Diffusers inference using TI

Before diving into how TI works internally, let's take a look at how to use TI using Diffusers.

There are countless pretrained TIs shared in the Hugging Face's Stable Diffusion concepts library [3] and CIVITAI [4]. For example, one of the most downloaded TIs from the Stable Diffusion concepts library is sd-concepts-library/midjourney-style [5]. We can start using it by simply referencing this name in the code; Diffusers will download the model data automatically:

1. Let's initialize a Stable Diffusion pipeline:

```
# initialize model
from diffusers import StableDiffusionPipeline
import torch

model_id = "stablediffusionapi/deliberate-v2"
pipe = StableDiffusionPipeline.from_pretrained(
    model_id,
    torch_dtype=torch.float16
).to("cuda")
```

2. Generate an image without TI involved:

```
# without using TI
prompt = "a high quality photo of a futuristic city in deep \
space, midjourney-style"

image = pipe(
    prompt,
    num_inference_steps = 50,
    generator = torch.Generator("cuda").manual_seed(1)
).images[0]
image
```

In the prompt, midjourney-style is used, which will be given as the name of the TI. Without applying the name, we will see an image generated, as shown in *Figure 9.1*:

Figure 9.1: A futuristic city in deep space without TI

3. Generate an image with TI involved.

    Now, let's load the TI into the Stable Diffusion pipeline and give it a name, `midjourney-style`, to represent the newly added embeddings:

    ```
    pipe.load_textual_inversion(
        "sd-concepts-library/midjourney-style",
        token = "midjourney-style"
    )
    ```

    The preceding code will download the TI automatically and then add it to the pipeline model. Execute the same prompt and pipeline again, and we will get a completely new image, as shown in *Figure 9.2*:

Figure 9.2: A futuristic city in deep space with TI

Yes, it looks and feels like an image generated by Midjourney, but it is actually generated by Stable Diffusion. The *"inversion"* in the name of the TI indicates that we can inverse any new name to the new embeddings, for example, if we give a new token a name such as `colorful-magic-style`:

```
pipe.load_textual_inversion(
    "sd-concepts-library/midjourney-style",
    token = "colorful-magic-style"
)
```

We will get the same image because we use `midjourney-style` as the name of the TI. This time, we "inverse" `colorful-magic-style` into the new embeddings. However, the `load_textual_inversion` function provided by Diffusers does not provide a `weight` parameter for users to load a TI with a certain weight. We will add the weighted TI in our own TI loader later in this chapter.

Before that, let's dig into the heart of TI and see how it works internally.

How TI works

Simply put, training a TI is finding a text embedding that matches the target image the best, such as its style, object, or face. The key is to find a new embedding that never existed in the current text encoder. As *Figure 9.3*, from its original paper [1], shows:

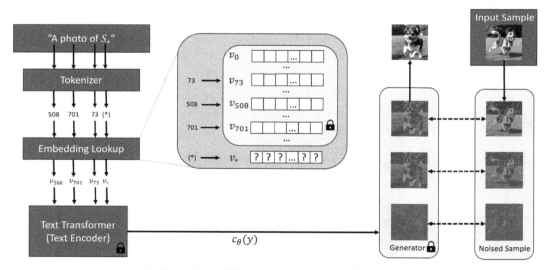

Figure 9.3: The outline of the text embedding and inversion process

The only job of the training is to find a new embedding represented by v_*. and use S_* as the token string placeholder; the string can be replaced by any string that does not exist in the tokenizer later. Once the new corresponding embedding vector is founded, the train is done. The output of the training is usually a vector with 768 numbers. That is why the TI file is tiny; it is just a couple of kilobytes.

It is like the pretrained UNet is a pile of matrix magic boxes, one key (embedding) can unlock a box to have a pattern, a style, or an object. The number of boxes is way more than the limited keys from the text encoder provided. The training of a TI is done by providing a new key to unlock the unknown magic box. Throughout the training and inferencing, the original checkpoint model is untouched.

In a precise way, the finding of the new embedding can be defined as follows:

$$v_* = \arg_v \min E_{z \sim E(x), y, \epsilon \sim N(0,1), t} \left[\left|\left| \epsilon - \epsilon_\theta(z_t, t, c_\theta(y)) \right|\right|_2^2 \right]$$

Let's go through the formula from left to right, one by one:

- v_* denotes the new embedding we are looking for.

- The $arg : min$ notation is often used in statistics and optimization to denote the set of values that minimize a function. It is a useful notation because it allows us to talk about the minimum value of a function without having to specify the actual value of the minimum.

- The $(\mathcal{E})$ is the loss expectation.

- $z \sim \mathcal{E}(x)$ denotes that the input image will be encoded to latent space.

- y is the input text.

- $e \sim \mathcal{N}(0,1)$ says that the initial noise latent is a strict Gaussian with 0 mean and 1 variance.

- $c_\theta(y)$ represents a text encoder model that maps an input text string y into embedding vectors.

- $\epsilon_\theta(z_t, t, c_\theta(y))$ means that we provide the noised latent image z in the t step, step t itself and text embeddings $c_\theta(y)$, and then generate noise vector from the UNet model.

- The 2 in $||^2$ means the square of the Euclidean distance. The 2 in $||_2$ means the data is in 2 dimensions.

Together, the formula shows how we can use Stable Diffusion's training process to approximate a new embedding, v_*, that generates the minimum loss.

Next, let's build a custom TI loader function.

Building a custom TI loader

In this section, we are going to build a TI loader by implementing the preceding understanding into code and giving a TI weight parameter for the loader function.

Before writing the function code, let's first take a look at how a TI looks internally. Before running the following code, you will need to first download the TI file to your storage.

TI in the pt file format

Load a TI in the pt file format:

```
# load a pt TI
import torch
loaded_learned_embeds = torch.load("badhandsv5-neg.pt",
    map_location="cpu")
keys = list(loaded_learned_embeds.keys())
for key in keys:
    print(key,":",loaded_learned_embeds[key])
```

We can clearly see the key and paired value from the TI file:

```
string_to_token : {'*': 265}
string_to_param : {'*': tensor([[ 0.0399,
-0.2473,  0.1252,  ...,  0.0455,  0.0845, -0.1463],
        [-0.1385, -0.0922, -0.0481,  ...,  0.1766, -0.1868,  0.3851]],
        requires_grad=True)}
name : bad-hands-5
step : 1364
sd_checkpoint : 7ab762a7
sd_checkpoint_name : blossom-extract
```

The most important value is the tensor object with the string_to_param key. We can take the tensor value out of it by using the following code:

```
string_to_token = loaded_learned_embeds['string_to_token']
string_to_param = loaded_learned_embeds['string_to_param']

# separate token and the embeds
trained_token = list(string_to_token.keys())[0]
embeds = string_to_param[trained_token]
```

TI in bin file format

Most of the TI from the Hugging Face concepts library is in the bin format. The bin structure is even simpler than the pt one:

```
import torch
loaded_learned_embeds = torch.load("midjourney_style.bin",
    map_location="cpu")
keys = list(loaded_learned_embeds.keys())
for key in keys:
    print(key,":",loaded_learned_embeds[key])
```

We will see this – a dictionary with just one key and one value:

```
<midjourney-style> : tensor([-5.9785e-02, -3.8523e-02,  5.1913e-
02,  8.0925e-03, -6.2018e-02,
        1.3361e-01,  1.3679e-01,  8.2224e-02, -2.0598e-01,  1.8543e-
02,
        1.9180e-01, -1.5537e-01, -1.5216e-01, -1.2607e-01, -1.9420e-
01,
        1.0445e-01,  1.6942e-01,  4.2150e-02, -2.7406e-01,  1.8115e-
01,
        ...
])
```

Extracting the tensor object is as simple as doing the following:

```
keys = list(loaded_learned_embeds.keys())
embeds =  loaded_learned_embeds[keys[0]] * weight
```

Detailed steps to build a TI loader

Here are the detailed steps to load a TI with weight:

1. **Load the embeddings**: We will reuse the code from the preceding two conditions, and add another condition that some TIs use the emb_params key to store the embedding tensor.

    Use this function to load a TI in the model initialization stage or image generation stage:

    ```
    def load_textual_inversion(
        learned_embeds_path,
        token,
        text_encoder,
        tokenizer,
        weight = 0.5,
        device = "cpu"
    ):
        loaded_learned_embeds = \
            torch.load(learned_embeds_path, map_location=device)
        if "string_to_token" in loaded_learned_embeds:
            string_to_token = \
                loaded_learned_embeds['string_to_token']
            string_to_param = \
                loaded_learned_embeds['string_to_param']

            # separate token and the embeds
            trained_token = list(string_to_token.keys())[0]
            embeds = string_to_param[trained_token]
    ```

```
        embeds = embeds[0] * weight
    elif "emb_params" in loaded_learned_embeds:
        embeds = loaded_learned_embeds["emb_params"][0] * weight
    else:
        keys = list(loaded_learned_embeds.keys())
        embeds =  loaded_learned_embeds[keys[0]] * weight
    # ...
```

Let's break down the preceding code:

- `torch.load(learned_embeds_path, map_location=device)` loads the learned embeddings from the specified file using PyTorch's `torch.load` function

- `if "string_to_token" in loaded_learned_embeds` then checks for a specific file structure where embeddings are stored in a dictionary with the `string_to_token` and `string_to_param` keys, and extracts the token and embeddings from this structure

- `elif "emb_params" in loaded_learned_embeds` then handles a different structure where embeddings are directly stored under the `emb_params` key

- `else:` then handles a generic structure by assuming the embeddings are stored under the first key of the dictionary

In essence, the weight serves as a multiplier for each element of the embedding vector, fine-tuning the intensity of the TI effect. For example, a weight value of `1.0` would apply the TI at full strength, while a value of `0.5` would apply it at half strength.

2. Cast data to the same type of Stable Diffusion text encoder:

```
dtype = text_encoder.get_input_embeddings().weight.dtype
embeds.to(dtype)
```

3. Add the token to the tokenizer:

```
token = token if token is not None else trained_token
num_added_tokens = tokenizer.add_tokens(token)
if num_added_tokens == 0:
    raise ValueError(
        f"""The tokenizer already contains the token {token}.
        Please pass a different `token` that is not already in
        the tokenizer."""
    )
```

The code will raise an exception if the added token already exists to prevent overriding existing tokens.

4.   Get the token ID and add the new embedding to the text encoder:

```
# resize the token embeddings
text_encoder.resize_token_embeddings(len(tokenizer))

# get the id for the token and assign the embeds
token_id = tokenizer.convert_tokens_to_ids(token)
text_encoder.get_input_embeddings().weight.data[token_id] = embeds
```

That is all the code needs to load most of the existing TI from both the Hugging Face and Civitai.

Putting all of the code together

Let's put all of the code blocks together into one function – load_textual_inversion:

```
def load_textual_inversion(
    learned_embeds_path,
    token,
    text_encoder,
    tokenizer,
    weight = 0.5,
    device = "cpu"
):
    loaded_learned_embeds = torch.load(learned_embeds_path,
        map_location=device)
    if "string_to_token" in loaded_learned_embeds:
        string_to_token = loaded_learned_embeds['string_to_token']
        string_to_param = loaded_learned_embeds['string_to_param']

        # separate token and the embeds
        trained_token = list(string_to_token.keys())[0]
        embeds = string_to_param[trained_token]
        embeds = embeds[0] * weight
    elif "emb_params" in loaded_learned_embeds:
        embeds = loaded_learned_embeds["emb_params"][0] * weight
    else:
        keys = list(loaded_learned_embeds.keys())
        embeds =  loaded_learned_embeds[keys[0]] * weight

    # cast to dtype of text_encoder
    dtype = text_encoder.get_input_embeddings().weight.dtype
    embeds.to(dtype)

    # add the token in tokenizer
```

```
token = token if token is not None else trained_token
num_added_tokens = tokenizer.add_tokens(token)
if num_added_tokens == 0:
    raise ValueError(
        f"""The tokenizer already contains the token {token}.
        Please pass a different `token` that is not already in the
        tokenizer."""
    )

# resize the token embeddings
text_encoder.resize_token_embeddings(len(tokenizer))

# get the id for the token and assign the embeds
token_id = tokenizer.convert_tokens_to_ids(token)
text_encoder.get_input_embeddings().weight.data[token_id] = embeds
return (tokenizer,text_encoder)
```

To use it, we need to have both `tokenizer` and `text_encoder` from the pipeline object:

```
text_encoder = pipe.text_encoder
tokenizer = pipe.tokenizer
```

Then load it by calling the newly created function:

```
load_textual_inversion(
    learned_embeds_path = "learned_embeds.bin",
    token = "colorful-magic-style",
    text_encoder = text_encoder,
    tokenizer = tokenizer,
    weight = 0.5,
    device = "cuda"
)
```

Now, use the same inference code to generate an image. Note that this time, we are using TI with a weight of 0.5, so, let's see whether anything is different compared with the original one with a weight of 1.0:

```
prompt = "a high quality photo of a futuristic city in deep space,
colorful-magic-style"

image = pipe(
    prompt,
    num_inference_steps = 50,
    generator = torch.Generator("cuda").manual_seed(1)
```

```
).images[0]
image
```

The result seems quite good (see *Figure 9.4*):

Figure 9.4: A futuristic city in deep space, with TI loaded by a custom function

The result seems even better than the one using Diffusers' one-line TI loader. Another advantage of our custom loader is that we can now freely give weight to the loaded model.

Summary

This chapter discussed what Stable Diffusion TI is and the difference between it and LoRA. Then, we introduced a quick way to load any TI into Diffusers to apply a new pattern, style, or object in the generation pipeline.

Then, we dove into the core of TI and learned about how it is trained and how it works. Based on the understanding of how it works, we went a step further to implement a TI loader with the capability of accepting a TI weight.

Lastly, we provided a piece of sample code to call the custom TI loader and then generate an image with a weight of 0.5.

In the next chapter, we'll explore ways to maximize the power of prompts and unlock their full potential.

References

1. Rinon et al., *An Image is Worth One Word: Personalizing Text-to-Image Generation using Textual Inversion*: `https://arxiv.org/abs/2208.01618` and `https://textual-inversion.github.io/`

2. Hugging Face, *Textual Inversion*: `https://huggingface.co/docs/diffusers/main/en/training/text_inversion#how-it-works`

3. *Stable Diffusion concepts library*: `https://huggingface.co/sd-concepts-library`

4. Civitai: `https://civitai.com`

5. *Midjourney style on Stable Diffusion*: `https://huggingface.co/sd-concepts-library/midjourney-style`

6. OpenAI's CLIP: `https://github.com/openai/CLIP`

10

Overcoming 77-Token Limitations and Enabling Prompt Weighting

From *Chapter 5*, we know that Stable Diffusion utilizes OpenAI's CLIP model as its text encoder. The CLIP model's tokenization implementation, as per the source code [6], has a context length of 77 tokens.

This 77-token limit in the CLIP model extends to Hugging Face Diffusers, restricting the maximum input prompt to 77 tokens. Unfortunately, it's not possible to assign keyword weights within these input prompts due to this constraint without some modifications.

For instance, let's say you give a prompt string that produces more than 77 tokens, like this:

```
from diffusers import StableDiffusionPipeline
import torch

pipe = StableDiffusionPipeline.from_pretrained(
    "stablediffusionapi/deliberate-v2",
    torch_dtype=torch.float16).to("cuda")

prompt = "a photo of a cat and a dog driving an aircraft "*20
image = pipe(prompt = prompt).images[0]
image
```

Diffusers will show up a warning message, like this:

```
The following part of your input was truncated because CLIP can only
handle sequences up to 77 tokens...
```

You can't highlight the cat by providing the weight, such as in the following:

```
a photo (cat:1.5) and a dog driving an aircraft
```

By default, the `Diffusers` package does not include functionality for overcoming the 77-token limitation or assigning weights to individual tokens, as stated in its documentation. This is because Diffusers aims to serve as a versatile toolbox, providing essential features that can be utilized in various projects.

Nonetheless, using the core functionalities offered by Diffusers, we can develop a custom prompt parser. This parser will help us circumvent the 77-token restriction and assign weights to each token. In this chapter, we will delve into the structure of text embeddings and outline a method to surpass the 77-token limit while also assigning weight values to each token.

In this chapter, we will cover the following:

- Understanding the 77-token limitation
- Overcoming the 77-token limitation
- Enabling long prompts with weighting
- Overcoming the 77-token limitation using community pipelines

If you want to start using a full feature pipeline supporting long prompt weighting, please turn to the *Overcoming the 77-token limitation using the community pipelines* section.

By the end of the chapter, you will be able to use weighted prompts without size limitations and know how to implement them using Python.

Understanding the 77-token limitation

The Stable Diffusion (v1.5) text encoder uses the CLIP encoder from OpenAI [2]. The CLIP text encoder has a 77-token limit, and this limitation propagates to the downstream Stable Diffusion. We can reproduce the 77-token limitation with the following steps:

1. We can take out the encoder from Stable Diffusion and verify it. Let's say we have the prompt `a photo of a cat and dog driving an aircraft` and we multiply it by 20 to make the prompt's token size larger than 77:

    ```
    prompt = "a photo of a cat and a dog driving an aircraft "*20
    ```

2. Reuse the pipeline we initialized at the beginning of the chapter and take out `tokenizer` and `text_encoder`:

    ```
    tokenizer = pipe.tokenizer
    text_encoder = pipe.text_encoder
    ```

3. Use `tokenizer` to get the token IDs from the prompt:

    ```
    tokens = tokenizer(
        prompt,
        truncation = False,
    ```

```
        return_tensors = 'pt'
    )["input_ids"]
    print(len(tokens[0]))
```

4. Since we set truncation = False, tokenizer will convert any length string to token IDs. The preceding code will output a token list with a length of 181. return_tensors = 'pt' will tell the function to return the result in a [1,181] tensor object.

    Try encoding the token IDs to embeddings:

    ```
    embeddings = pipe.text_encoder(tokens.to("cuda"))[0]
    ```

    We will see a RuntimeError message that says the following:

    ```
    RuntimeError: The size of tensor a (181) must match the size of
    tensor b (77) at non-singleton dimension 1
    ```

    From the preceding steps, we can see that CLIP's text encoder only accepts 77 tokens at a time.

5. Now, let's take a look at the first and last tokens. If we take away *20 and only tokenize the prompt a photo cat and dog driving an aircraft, when we print out the token IDs, we will see 10 token IDs instead of 8:

    ```
    tensor([49406,   320,  1125,  2368,   537,  1929,  4161,   550
    ,  7706, 49407])
    ```

6. In the preceding token IDs, the first (49406) and last (49407) are automatically added. We can use tokenizer._convert_id_to_token to convert the token IDs to a string:

    ```
    print(tokenizer._convert_id_to_token(49406))
    print(tokenizer._convert_id_to_token(49407))
    ```

    We can see that the two additional tokens are added to the prompt:

    ```
    <|startoftext|>
    <|endoftext|>
    ```

Why do we need to check this? Because we need to remove the automatically added beginning and end tokens when concatenating tokens. Next, let's proceed to the steps to overcome the 77-token limitation.

Overcoming the 77-tokens limitation

Fortunately, the Stable Diffusion UNet doesn't enforce this 77-token limitation. If we can get the embeddings in batches, concatenate those chunked embeddings into one tensor, and provide it to the UNet, we should be able to overcome the 77-token limitation. Here's an overview of the process:

1. Extract the text tokenizer and text encoder from the Stable Diffusion pipeline.

2. Tokenize the input prompt, regardless of its size.

3. Eliminate the added beginning and end tokens.

4. Pop out the first 77 tokens and encode them into embeddings.

5. Stack the embeddings into a tensor of size `[1, x, 768]`.

Now, let's implement this idea using Python code:

1. Take out the text tokenizer and text encoder:

```
# step 1. take out the tokenizer and text encoder
tokenizer = pipe.tokenizer
text_encoder = pipe.text_encoder
```

We can reuse the tokenizer and text encoder from the Stable Diffusion pipeline.

2. Tokenize any size of input prompt:

```
# step 2. encode whatever size prompt to tokens by setting
# truncation = False.
tokens = tokenizer(
    prompt,
    truncation = False
)["input_ids"]
print("token length:", len(tokens))

# step 2.2. encode whatever size neg_prompt,
# padding it to the size of prompt.
negative_ids = pipe.tokenizer(
    neg_prompt,
    truncation     = False,
    padding        = "max_length",
    max_length     = len(tokens)
).input_ids
print("neg_token length:", len(negative_ids))
```

In the preceding code, we did the following:

- We set `truncation = False` to allow tokenization beyond the default 77-token limit. This ensures that the entire prompt is tokenized, regardless of its size.

- The tokens are returned as a Python list instead of a torch tensor. Tokens in the Python list will make it easier for us to add additional elements. Note that the token list will be converted to a torch tensor before providing it to the text encoder.

- There are two additional parameters, `padding = "max_length"` and `max_length = len(tokens)`. We use these to make sure prompt tokens and negative prompt tokens are the same size.

3. Remove the beginning and end tokens.

The tokenizer will automatically add two additional tokens: the beginning token (49406) and the end token (49407).

In the subsequent step, we will segment the token sequence and feed the chunked tokens to the text encoder. Each chunk will have its own beginning and end tokens. But before that, we will need to exclude them initially from the long original token list:

```
tokens = tokens[1:-1]
negative_ids = negative_ids[1:-1]
```

And then add these beginning and end tokens back to the chunked tokens, each chunk with a size of size 75. We will add the beginning and the end tokens back in *step 4*.

4. Encode the 77-sized chunked tokens into embeddings:

```
# step 4. Pop out the head 77 tokens,
# and encode the 77 tokens to embeddings.
embeds,neg_embeds = [],[]
chunk_size = 75
bos = pipe.tokenizer.bos_token_id
eos = pipe.tokenizer.eos_token_id
for i in range(0, len(tokens), chunk_size):
# Add the beginning and end token to the 75 chunked tokens to
# make a 77-token list
    sub_tokens = [bos] + tokens[i:i + chunk_size] + [eos]

# text_encoder support torch.Size([1,x]) input tensor
# that is why use [sub_tokens],
# instead of simply give sub_tokens.
    tensor_tokens = torch.tensor(
        [sub_tokens],
        dtype = torch.long,
        device = pipe.device
    )
    chunk_embeds = text_encoder(tensor_tokens)[0]
    embeds.append(chunk_embeds)

# Add the begin and end token to the 75 chunked neg tokens to
# make a 77 token list
    sub_neg_tokens = [bos] + negative_ids[i:i + chunk_size] + \
        [eos]
    tensor_neg_tokens = torch.tensor(
        [sub_neg_tokens],
        dtype = torch.long,
        device = pipe.device
```

```
)
neg_chunk_embeds= text_encoder(tensor_neg_tokens)[0]
neg_embeds.append(neg_chunk_embeds)
```

The preceding code loops through the token list, taking out 75 tokens at a time. Then, it adds the beginning and end tokens to the 75-token list to create a 77-token list. Why 77 tokens? Because the text encoder can encode 77 tokens to embeddings at one time.

Inside the `for` loop, the first part handles the prompt embeddings, and the second part handles the negative embeddings. Even though we give an empty negative prompt, to enable classification-free guidance diffusion, we still need a negative embedding list with the same size of positive prompt embeddings (inside of the denoising loop, the conditioned latent will subtract the unconditional latent, which is generated from the negative prompt).

5. Stack the embeddings to a `[1,x,768]` size torch tensor.

    Before this step, the `embeds` list holds data like this:

    ```
    [tensor1, tensor2...]
    ```

    The Stable Diffusion pipeline's embedding parameters accept tensors in the size of `torch.Size([1,x,768])`.

    We still need to convert the list to a three-dimension tensor using these two lines of code:

    ```
    # step 5. Stack the embeddings to a [1,x,768] size torch tensor.
    prompt_embeds = torch.cat(embeds, dim = 1)
    prompt_neg_embeds = torch.cat(neg_embeds, dim = 1)
    ```

    In the preceding code, we have the following:

    *   `embeds` and `neg_embeds` are lists of PyTorch tensors. The `torch.cat()` function is used to concatenate these tensors along the dimension specified by `dim`. In this case, we have `dim=1`, which means the tensors are concatenated along their second dimension (since Python uses 0-based indexing).

    *   `prompt_embeds` is a tensor that contains all the embeddings from `embeds` concatenated together. Similarly, `prompt_neg_embeds` contains all the embeddings from `neg_embeds` concatenated together.

By now, we have a functioning text encoder that can convert whatever length of prompt to embeddings, which can be used by a Stable Diffusion pipeline. Next, let's put all the code together.

Putting all the code together into a function

Let's go a step further to put all the previous code in a packed function:

```
def long_prompt_encoding(
    pipe:StableDiffusionPipeline,
```

```
    prompt,
    neg_prompt = ""
):
    bos = pipe.tokenizer.bos_token_id
    eos = pipe.tokenizer.eos_token_id
    chunk_size = 75

    # step 1. take out the tokenizer and text encoder
    tokenizer = pipe.tokenizer
    text_encoder = pipe.text_encoder

    # step 2.1. encode whatever size prompt to tokens by setting
    # truncation = False.
    tokens = tokenizer(
        prompt.
        truncation = False,
        # return_tensors = 'pt'
    )["input_ids"]

    # step 2.2. encode whatever size neg_prompt,
    # padding it to the size of prompt.
    negative_ids = pipe.tokenizer(
        neg_prompt,
        truncation = False,
        # return_tensors = "pt",
        Padding = "max_length",
        max_length = len(tokens)
    ).input_ids

    # Step 3. remove begin and end tokens
    tokens = tokens[1:-1]
    negative_ids = negative_ids[1:-1]

    # step 4. Pop out the head 77 tokens,
    # and encode the 77 tokens to embeddings.
    embeds,neg_embeds = [],[]
    for i in range(0, len(tokens), chunk_size):
# Add the beginning and end tokens to the 75 chunked tokens to make a
# 77-token list
        sub_tokens = [bos] + tokens[i:i + chunk_size] + [eos]

# text_encoder support torch.Size([1,x]) input tensor
# that is why use [sub_tokens], instead of simply give sub_tokens.
```

```
        tensor_tokens = torch.tensor(
            [sub_tokens],
            dtype = torch.long,
            device = pipe.device
        )
        chunk_embeds = text_encoder(tensor_tokens)[0]
        embeds.append(chunk_embeds)

    # Add beginning and end token to the 75 chunked neg tokens to make a
    # 77-token list
        sub_neg_tokens = [bos] + negative_ids[i:i + chunk_size] + \
            [eos]
        tensor_neg_tokens = torch.tensor(
            [sub_neg_tokens],
            dtype = torch.long,
            device = pipe.device
        )
        neg_chunk_embeds = text_encoder(tensor_neg_tokens)[0]
        neg_embeds.append(neg_chunk_embeds)

    # step 5. Stack the embeddings to a [1,x,768] size torch tensor.
        prompt_embeds = torch.cat(embeds, dim = 1)
        prompt_neg_embeds = torch.cat(neg_embeds, dim = 1)

        return prompt_embeds, prompt_neg_embeds
```

Let's create a long prompt to test whether the preceding function works or not:

```
prompt = "photo, cute cat running on the grass" * 10 #<- long prompt
prompt_embeds, prompt_neg_embeds = long_prompt_encoding(
    pipe, prompt, neg_prompt="low resolution, bad anatomy"
)
print(prompt_embeds.shape)

image = pipe(
    prompt_embeds = prompt_embeds,
    negative_prompt_embeds = prompt_neg_embeds,
    generator = torch.Generator("cuda").manual_seed(1)
).images[0]
image
```

The result is shown in *Figure 10.1*:

Figure 10.1: Cute cat running on the grass, using a long prompt

If our new function works for long prompts, the generated image should reflect additional appended prompts. Let's extend the prompt to the following:

```
prompt = "photo, cute cat running on the grass" * 10
prompt = prompt + ",pure white cat" * 10
```

The new prompt will generate an image as shown in *Figure 10.2*:

Figure 10.2: Cute cat running on the grass, with the additional prompt of pure white cat

As you can see, the new appended prompt works and there are more white elements added to the cat; however, it is still not pure white as requested in the prompt. We will solve this with prompt weighting, which we'll cover in the upcoming section.

Enabling long prompts with weighting

We just built a whatever size of text encoder for a Stable Diffusion pipeline (v1.5-based). All of those steps are paving the way to build long prompts with a weighting text encoder.

A weighted Stable Diffusion prompt refers to the practice of assigning different levels of importance to specific words or phrases within a text prompt used for generating images through the Stable Diffusion algorithm. By adjusting these weights, we can control the degree to which certain concepts influence the generated output, allowing for greater customization and refinement of the resulting images.

The process typically involves scaling up or down the text embedding vectors associated with each concept in the prompt. For instance, if you want the Stable Diffusion model to emphasize a particular subject while deemphasizing another, you would increase the weight of the former and decrease the weight of the latter. Weighted prompts enable us to better direct the image generation toward desired outcomes.

The core of adding weight to the prompt is simply vector multiplication:

$$weighted\_embeddings = [embedding1, embedding2, ..., embedding768] \times weight$$

Before that, we still need to do some preparations to make a weighted prompt embedding, as follows:

1. **Prompt parsing**: Parse the prompt string to extract weight numbers. For instance, convert the prompt `a (white) cat` into a list like this: `[['a ', 1.0], ['white', 1.1], ['cat', 1.0]]`. We'll adopt the prevalent prompt format used in the Automatic1111 **Stable Diffusion (SD)** WebUI, as defined in the open source SD WebUI [4].

2. **Token and weight extraction**: Separate the token IDs and their corresponding weights into two distinct lists.

3. **Prompt and negative prompt padding**: Ensure that both the prompt and negative prompt tokens have the same maximum length. If the prompt is longer than the negative prompt, pad the negative prompt to match the prompt's length. Otherwise, pad the prompt to align with the negative prompt's length.

Regarding attention and emphasis (weighting), we will implement the following weighting format [4]:

```
a (word) - increase attention to word by a factor of 1.1
a ((word)) - increase attention to word by a factor of 1.21 (= 1.1 *
1.1)
a [word] - decrease attention to word by a factor of 1.1
a (word:1.5) - increase attention to word by a factor of 1.5
a (word:0.25) - decrease attention to word by a factor of 4 (= 1 /
0.25)
a \(word\) - use literal () characters in prompt
```

Let's go through each of those steps in more detail:

1. Build the `parse_prompt_attention` function.

    To make sure the prompt format is fully compatible with Automatic1111's SD WebUI, we will extract and reuse the function from the open sourced `parse_prompt_attention` function [3]:

```
def parse_prompt_attention(text):
    import re
    re_attention = re.compile(
        r"""
        \\\(|\\\)|\\\[|\\\]|\\\\|\\|\(|\[|:([+-]?[.\d]+)\)|
        \)|\]|[^\\()\[\]:]+|:
        """
        , re.X
    )

    re_break = re.compile(r"\s*\bBREAK\b\s*", re.S)

    res = []
    round_brackets = []
    square_brackets = []

    round_bracket_multiplier = 1.1
    square_bracket_multiplier = 1 / 1.1

    def multiply_range(start_position, multiplier):
        for p in range(start_position, len(res)):
            res[p][1] *= multiplier

    for m in re_attention.finditer(text):
        text = m.group(0)
        weight = m.group(1)

        if text.startswith('\\'):
            res.append([text[1:], 1.0])
        elif text == '(':
            round_brackets.append(len(res))
        elif text == '[':
            square_brackets.append(len(res))
        elif weight is not None and len(round_brackets) > 0:
            multiply_range(round_brackets.pop(), float(weight))
        elif text == ')' and len(round_brackets) > 0:
            multiply_range(round_brackets.pop(), \
```

```
                        round_bracket_multiplier)
            elif text == ']' and len(square_brackets) > 0:
                multiply_range(square_brackets.pop(), \
                    square_bracket_multiplier)
            else:
                parts = re.split(re_break, text)
                for i, part in enumerate(parts):
                    if i > 0:
                        res.append(["BREAK", -1])
                    res.append([part, 1.0])

        for pos in round_brackets:
            multiply_range(pos, round_bracket_multiplier)

        for pos in square_brackets:
            multiply_range(pos, square_bracket_multiplier)

        if len(res) == 0:
            res = [["", 1.0]]

        # merge runs of identical weights
        i = 0
        while i + 1 < len(res):
            if res[i][1] == res[i + 1][1]:
                res[i][0] += res[i + 1][0]
                res.pop(i + 1)
            else:
                i += 1
        return res
```

Call the previously created function using the following:

```
parse_prompt_attention("a (white) cat")
```

This will return the following:

```
[['a ', 1.0], ['white', 1.1], [' cat', 1.0]]
```

2. Get prompts with weights.

    With the help of the preceding function, we can get a list of prompt and weight pairs. The text encoder will only encode the tokens of the prompt (no, you don't need to provide weights as input to the text encoder). We will need to further process the prompt-weight pair to two independent lists with the same size, one for token IDs and one for weights, like this:

    ```
    tokens: [1,2,3...]
    weights: [1.0, 1.0, 1.0...]
    ```

This work can be done by the following function:

```
# step 2. get prompts with weights
# this function works for both prompt and negative prompt
def get_prompts_tokens_with_weights(
    pipe: StableDiffusionPipeline,
    prompt: str
):
    texts_and_weights = parse_prompt_attention(prompt)
    text_tokens,text_weights = [],[]
    for word, weight in texts_and_weights:
        # tokenize and discard the starting and the ending token
        token = pipe.tokenizer(
            word,
            # so that tokenize whatever length prompt
            truncation = False
        ).input_ids[1:-1]
        # the returned token is a 1d list: [320, 1125, 539, 320]

        # use merge the new tokens to the all tokens holder:
        # text_tokens
        text_tokens = [*text_tokens,*token]

        # each token chunk will come with one weight, like ['red
        # cat', 2.0]
        # need to expand the weight for each token.
        chunk_weights = [weight] * len(token)

        # append the weight back to the weight holder: text_
        # weights
        text_weights = [*text_weights, *chunk_weights]
    return text_tokens,text_weights
```

The preceding function takes two parameters: the SD pipeline and the prompt string. The input string can be the positive prompt or the negative prompt.

Inside the function body, we first call the parse_prompt_attention function to have the prompts with weight associated in the smallest grain (The weight is applied in the individual token level). Then, we loop through the list, tokenize the text, and remove the tokenizer-added beginning and end token IDs with the indexing operation, [1:-1].

Merge the new token IDs back to the list that holds all token IDs. In the meantime, expand the weights number and merge back to the list that holds all weights numbers.

Let's reuse the prompt of a (white) cat and call the function:

```
prompt = "a (white) cat"
tokens, weights = get_prompts_tokens_with_weights(pipe, prompt)
print(tokens,weights)
```

The preceding code will return the following:

```
[320, 1579, 2368] [1.0, 1.1, 1.0]
```

Notice that the second token ID from white now has a weight of 1.1 instead of 1.0.

3. Pad the tokens.

    In this step, we will further transform the token ID list and its weights into a chunked list.

    Let's say we have a list of token IDs containing more than 77 elements:

    ```
    [1,2,3,...,100]
    ```

    We need to transform it into a list that includes chunks, with 77 (maximum) tokens inside each chunk:

    ```
    [[49406,1,2...75,49407],[49406,76,77,...,100,49407]]
    ```

    This is so that, in the next step, we can loop through the outer layer of the list and encode the 77-token list one at a time.

    Now, you may wonder why we need to provide a maximum of 77 tokens to the text encoder at a time. What if we simply loop through each element and encode one token at a time? Good question, but we can't do it like this because encoding white and then encoding cat will produce different embeddings compared with encoding white cat together at one time.

    We can use a quick test to find out the difference. First, let's encode white only:

    ```
    # encode "white" only
    white_token = 1579
    white_token_tensor = torch.tensor(
        [[white_token]],
        dtype = torch.long,
        device = pipe.device
    )
    white_embed = pipe.text_encoder(white_token_tensor)[0]
    print(white_embed[0][0])
    ```

    Then, encode white cat together:

    ```
    # encode "white cat"
    white_token, cat_token = 1579, 2369
    white_cat_token_tensor = torch.tensor(
        [[white_token, cat_token]],
        dtype = torch.long,
        device = pipe.device
    ```

```
)
white_cat_embeds = pipe.text_encoder(white_cat_token_tensor)[0]
print(white_cat_embeds[0][0])
```

Give the preceding code a try; you will find that the same `white` will lead to a different embedding. What is the root cause? The token and embedding is not a one-to-one mapping; the embedding is generated based on the self-attention mechanism [5]. A single `white` can represent the color or a family name, while `white` in `white cat` is clearly saying that it is a color that describes the cat.

Let's get back to the padding work. The following code will check the length of the token list. If the token ID list length is larger than 75, then take the first 75 tokens and loop this operation the remaining tokens are fewer than 75, which will be handled by a separate logic:

```
# step 3. padding tokens
def pad_tokens_and_weights(
    token_ids: list,
    weights: list
):
    bos,eos = 49406,49407

    # this will be a 2d list
    new_token_ids = []
    new_weights   = []
    while len(token_ids) >= 75:
        # get the first 75 tokens
        head_75_tokens = [token_ids.pop(0) for _ in range(75)]
        head_75_weights = [weights.pop(0) for _ in range(75)]

        # extract token ids and weights
        temp_77_token_ids = [bos] + head_75_tokens + [eos]
        temp_77_weights   = [1.0] + head_75_weights + [1.0]

        # add 77 tokens and weights chunks to the holder list
        new_token_ids.append(temp_77_token_ids)
        new_weights.append(temp_77_weights)

    # padding the left
    if len(token_ids) > 0:
        padding_len = 75 - len(token_ids)
        padding_len = 0

        temp_77_token_ids = [bos] + token_ids + [eos] * \
            padding_len + [eos]
        new_token_ids.append(temp_77_token_ids)
```

```
            temp_77_weights = [1.0] + weights   + [1.0] * \
                padding_len + [1.0]
            new_weights.append(temp_77_weights)

        # return
        return new_token_ids, new_weights
```

Next, use the following function:

```
t,w = pad_tokens_and_weights(tokens.copy(), weights.copy())
print(t)
print(w)
```

The preceding function takes the following previously generated `tokens` and `weights` list:

```
[320, 1579, 2368] [1.0, 1.1, 1.0]
```

It transforms it into this:

```
[[49406, 320, 1579, 2368, 49407]]
[[1.0, 1.0, 1.1, 1.0, 1.0]]
```

4. Get the weighted embeddings.

This is the final step, and we will get the Automatic1111-compatible embeddings without a token size limitation:

```
def get_weighted_text_embeddings(
    pipe: StableDiffusionPipeline,
    prompt : str        = "",
    neg_prompt: str     = ""
):
    eos = pipe.tokenizer.eos_token_id
    prompt_tokens, prompt_weights = \
        get_prompts_tokens_with_weights(
        pipe, prompt
    )
    neg_prompt_tokens, neg_prompt_weights = \
        get_prompts_tokens_with_weights(pipe, neg_prompt)

    # padding the shorter one
    prompt_token_len        = len(prompt_tokens)
    neg_prompt_token_len    = len(neg_prompt_tokens)
    if prompt_token_len > neg_prompt_token_len:
        # padding the neg_prompt with eos token
        neg_prompt_tokens    = (
            neg_prompt_tokens  + \
```

```
            [eos] * abs(prompt_token_len - neg_prompt_token_len)
        )
        neg_prompt_weights  = (
            neg_prompt_weights +
            [1.0] * abs(prompt_token_len - neg_prompt_token_len)
        )
    else:
        # padding the prompt
        prompt_tokens        = (
            prompt_tokens \
            + [eos] * abs(prompt_token_len - \
            neg_prompt_token_len)
        )
        prompt_weights       = (
            prompt_weights \
            + [1.0] * abs(prompt_token_len - \
            neg_prompt_token_len)
        )

embeds = []
neg_embeds = []

prompt_token_groups ,prompt_weight_groups = \
    pad_tokens_and_weights(
        prompt_tokens.copy(),
        prompt_weights.copy()
)

neg_prompt_token_groups, neg_prompt_weight_groups = \
    pad_tokens_and_weights(
        neg_prompt_tokens.copy(),
        neg_prompt_weights.copy()
    )

# get prompt embeddings one by one is not working.
for i in range(len(prompt_token_groups)):
    # get positive prompt embeddings with weights
    token_tensor = torch.tensor(
        [prompt_token_groups[i]],
        dtype = torch.long, device = pipe.device
    )
    weight_tensor = torch.tensor(
        prompt_weight_groups[i],
```

```
        dtype     = torch.float16,
        device    = pipe.device
    )
    token_embedding = \
        pipe.text_encoder(token_tensor)[0].squeeze(0)
    for j in range(len(weight_tensor)):
        token_embedding[j] = token_embedding[j] *
            weight_tensor[j]
    token_embedding = token_embedding.unsqueeze(0)
    embeds.append(token_embedding)

    # get negative prompt embeddings with weights
    neg_token_tensor = torch.tensor(
        [neg_prompt_token_groups[i]],
        dtype = torch.long, device = pipe.device
    )
    neg_weight_tensor = torch.tensor(
        neg_prompt_weight_groups[i],
        dtype     = torch.float16,
        device    = pipe.device
    )
    neg_token_embedding = \
        pipe.text_encoder(neg_token_tensor)[0].squeeze(0)
    for z in range(len(neg_weight_tensor)):
        neg_token_embedding[z] = (
            neg_token_embedding[z] * neg_weight_tensor[z]
        )
    neg_token_embedding = neg_token_embedding.unsqueeze(0)
    neg_embeds.append(neg_token_embedding)

prompt_embeds      = torch.cat(embeds, dim = 1)
neg_prompt_embeds  = torch.cat(neg_embeds, dim = 1)

return prompt_embeds, neg_prompt_embeds
```

The function looks a bit long but the logic is simple. Let me explain it section by section:

- In the *padding-the-shorter-one* section, the logic will fill the shorter prompt with the ending token (eos) so that both the prompt and negative prompt token lists share the same size (so that the generated latent can do a subtraction operation).

- We call the pad_tokens_and_weights function to break all tokens and weights into chunks, each chunk with 77 elements.

- We loop through the chunk list and encode the 77 tokens to embed in one step.

- We use `token_embedding = pipe.text_encoder(token_tensor)[0].squeeze(0)` to remove empty dimensions, so that we can multiply each element with its weight. Note that, now, each token is represented by a 768-element vector.

- Finally, we exit the loop and stack the tensor list to a higher dimension tensor using `prompt_embeds = torch.cat(embeds, dim = 1)`.

Verifying the work

After the not-so-many lines of code, we finally have all the logic ready, so let's give the code a test.

In the simple version of a *long prompt encoder*, we still get a cat with some patterns in the body instead of `pure white`, as we gave in the prompt. Now, let's add weight to the `white` keyword to see whether anything happens:

```
prompt = "photo, cute cat running on the grass" * 10
prompt = prompt + ",pure (white:1.5) cat" * 10

neg_prompt = "low resolution, bad anatomy"

prompt_embeds, prompt_neg_embeds = get_weighted_text_embeddings(
    pipe, prompt = prompt, neg_prompt = neg_prompt
)

image = pipe(
    prompt_embeds = prompt_embeds,
    negative_prompt_embeds = prompt_neg_embeds,
    generator = torch.Generator("cuda").manual_seed(1)
).images[0]
image
```

Our new embedding function magically enabled us to generate a pure white cat, as we gave a `1.5` weight to the `white` keyword.

Figure 10.3: Cute pure white cat running on the grass, with a 1.5 weight on the word white

That is all! Now, we can reuse or extend this function to build any custom prompt parser as we want. But what if you don't want to build your own function to implement; are there ways to start using unlimited weighted prompts? Yes, next we are going to introduce two pipelines contributed from the open source community and integrated into Diffusers.

Overcoming the 77-token limitation using community pipelines

Implementing a pipeline supporting long prompt weighting from scratch can be challenging. Often, we simply wish to utilize Diffusers to generate images using detailed and nuanced prompts. Fortunately, the open source community has provided implementations for SD v1.5 and SDXL. The SDXL implementation was originally initialized by Andrew Zhu, the author of this book, and massively improved by the community.

I'll now provide two examples of how to use the community pipeline for SD v1.5 and SDXL:

1. This example uses the `lpw_stable_diffusion` pipeline for SD v1.5.

    Use the following code to start a long prompt weighted pipeline:

    ```
    from diffusers import DiffusionPipeline
    import torch

    model_id_or_path = "stablediffusionapi/deliberate-v2"
    pipe = DiffusionPipeline.from_pretrained(
        model_id_or_path,
        torch_dtype = torch.float16,
        custom_pipeline = "lpw_stable_diffusion"
    ).to("cuda:0")
    ```

    In the preceding code, `custom_pipeline = "lpw_stable_diffusion"` will actually download the `lpw_stable_diffusion` file from the Hugging Face server and will be invoked inside of the `DiffusionPipeline` pipeline.

2. Let's generate an image using the pipeline:

    ```
    prompt = "photo, cute cat running on the grass" * 10
    prompt = prompt + ",pure (white:1.5) cat" * 10

    neg_prompt = "low resolution, bad anatomy"
    image = pipe(
        prompt = prompt,
        negative_prompt = neg_prompt,
        generator = torch.Generator("cuda").manual_seed(1)
    ).images[0]
    image
    ```

You will see an image the same as in *Figure 10.3*.

3. Now let's see an example using the `lpw_stable_diffusion` pipeline for SDXL.

    The usage is almost the same as the one we used in SD v1.5. The only differences are that we are loading an SDXL model and that we use another custom pipeline name: `lpw_stable_diffusion_xl`. See the following code:

    ```
    from diffusers import DiffusionPipeline
    import torch

    model_id_or_path = "stabilityai/stable-diffusion-xl-base-1.0"
    pipe = DiffusionPipeline.from_pretrained(
        model_id_or_path,
        torch_dtype = torch.float16,
        custom_pipeline = "lpw_stable_diffusion_xl",
    ).to("cuda:0")
    ```

    The image generation code is exactly the same as the one we used for SD v1.5:

    ```
    prompt = "photo, cute cat running on the grass" * 10
    prompt = prompt + ",pure (white:1.5) cat" * 10

    neg_prompt = "low resolution, bad anatomy"
    image = pipe(
        prompt = prompt,
        negative_prompt = neg_prompt,
        generator = torch.Generator("cuda").manual_seed(7)
    ).images[0]
    image
    ```

We will see an image generated as shown in *Figure 10.4*:

Figure 10.4: Cute pure white cat running on the grass, with a 1.5 weight
on the word white, using lpw_stable_diffusion_xl

From the image, we can clearly see what `pure (white:1.5) cat` is bringing into the image: proof that the pipeline can be used to generate images using long weighted prompts.

Summary

This chapter tried to solve one of the most discussed topics: overcoming the 77-token limitation and adding prompt weights for the Stable Diffusion pipeline using the `Diffusers` package. Automatic1111's Stable Diffusion WebUI provides a versatile UI and is now (as I am writing this) the most prevailing prompt weighting and attention format. However, if we take a look at the code from Automatic1111, we will probably get lost soon; its code is long without clear documentation.

This chapter started with understanding the root cause of the 77-token limitation and advanced to how the Stable Diffusion pipeline uses prompt embeddings. We implemented two functions to overcome the 77-token limitation.

One simple function without weighting was implemented to show how to walk around the 77-token limitation. We also built another function with the full function of a long prompt usage without length limitations and also have prompt weighting implemented.

By understanding and implementing these two functions, we can leverage the idea to not only use Diffuser to produce high-quality images the same as we can by using Automatic1111's WebUI but we can also further extend it to add more powerful features. In terms of which feature to add, it is in your hands now. In the next chapter, we'll start another exciting topic: using Stable Diffusion to fix and upscale images.

References

1. Hugging Face, weighted prompts: `https://huggingface.co/docs/diffusers/main/en/using-diffusers/weighted_prompts`

2. OpenAI CLIP, Connecting text and images: `https://openai.com/research/clip`

3. Automatic1111, Stable Diffusion WebUI prompt parser: `https://github.com/AUTOMATIC1111/stable-diffusion-webui/blob/master/modules/prompt_parser.py#L345C19-L345C19`

4. Automatic1111, Attention/emphasis: `https://github.com/AUTOMATIC1111/stable-diffusion-webui/wiki/Features#attentionemphasis`

5. Ashish et al., *Attention Is All You Need*: `https://arxiv.org/abs/1706.03762`

6. Source of the 77-token size limitation: `https://github.com/openai/CLIP/blob/4d120f3ec35b30bd0f992f5d8af2d793aad98d2a/clip/clip.py#L206`

11

Image Restore and Super-Resolution

While both Stable Diffusion V1.5 and Stable Diffusion XL have demonstrated abilities in generating images, our initial creations might not yet exhibit their utmost quality. This chapter aims to explore various techniques and strategies that can elevate image restoration, amplify image resolution, and introduce intricate details to produced visuals.

The primary focus of this chapter lies in leveraging the potential of Stable Diffusion as an effective tool to enhance and upscale images. Furthermore, we will briefly introduce you to complementary cutting-edge **Artificial Intelligence** (**AI**) methodologies to boost image resolutions, which are distinct from diffusion-based processes.

In this chapter, we will cover the following topics:

- Understanding the terminologies
- Upscaling images using an image-to-image diffusion pipeline
- Upscaling images using ControlNet Tile

Let's start!

Understanding the terminologies

Before we begin using Stable Diffusion to enhance image quality, it is beneficial to understand the common terminologies associated with this process. Three related terms you may encounter in articles or books on Stable Diffusion are **image interpolation**, **image upscale**, and **image super-resolution**. These techniques aim to improve the resolution of an image, but they differ in their methods and outcomes. Familiarizing yourself with these terms will help you better understand how Stable Diffusion and other image enhancement tools work:

- **Image interpolation** stands out as the simplest and most prevalent method to upscale images. It functions by approximating new pixel values based on the existing pixels within an image. Various interpolation methods exist, each with its strengths and weaknesses. Some of the most commonly used interpolation methods include nearest-neighbor interpolation [6], bilinear interpolation [7], bicubic interpolation [8], and Lanczos resampling [9].

- **Image upscale** is a broader term encompassing any technique that augments an image's resolution. This category includes interpolation methods, as well as more sophisticated approaches such as super-resolution.

- **Image super-resolution** represents a specific subset of image upscaling aimed at enhancing an image's resolution and finer details beyond its original dimensions while minimizing quality loss and preventing artifacts. In contrast to conventional image upscaling methods, which rely on basic interpolation, image super-resolution employs advanced algorithms, often based on deep learning techniques. These algorithms learn high-frequency patterns and details from a dataset of high-resolution images. Subsequently, these learned patterns are employed to upscale low-resolution images, producing superior-quality results.

A solution to the aforementioned tasks (image interpolation, image upscale, and image super-resolution) is commonly referred to as an **upscaler** or a **high-res fixer**. We will use the term *upscaler* throughout the chapter.

Among the deep learning-based solutions for super-resolution, various types of upscalers exist. Super-resolution solutions can be broadly classified into three types:

- GAN-based solutions such as ESRGAN [10]
- Swin Transformer-based solutions such as SwinIR [11],
- Stable Diffusion-based solutions

In this chapter, our primary focus will be on the Stable Diffusion upscaler. This choice is driven not only by the book's emphasis on the diffusion model but also by the potential of Stable Diffusion to provide enhanced upscale results and superior control flexibility. For instance, it allows us to direct the super-resolution process with prompts and fill in additional details. We will implement this using Python code in the latter part of this chapter.

You might be eager to start utilizing the Latent Upscaler and Stable Diffusion Upscale pipeline [1] from Diffusers. However, it's worth noting that the current Latent Upscaler and Upscale pipeline are not optimal. Both rely heavily on a specific pre-trained model, consume a substantial amount of VRAM, and exhibit relatively slower performance.

This chapter will introduce two alternative solutions based on the Stable Diffusion approach:

- **Image super-resolution using the img-to-img pipeline**: This approach enables the use of any SD (v1.5 and SDXL) based models and even permits the integration of LoRA to assist in image upscaling. Additionally, this solution is built upon the foundational `StableDiffusionPipeline` class. It also retains support for lengthy prompts and incorporates Textual Inversion, as introduced in previous chapters.

- **ControlNet-based Tile super-resolution**: This solution also offers compatibility with arbitrary pre-trained models. Through the incorporation of an additional `ControlNet` model, it becomes possible to achieve image super-resolution with remarkable enhancements in detail.

With this background, let's delve into the intricate details of these two super-resolution methods.

Upscaling images using Img2img diffusion

As we discussed in *Chapter 5*, Stable Diffusion doesn't solely rely on text as its initial guidance; it is also capable of utilizing an image as the starting point. We implemented a custom pipeline that employs an image as the foundation for image generation.

By reducing the denoising strength to a certain threshold, such as 0.3, the features and style of the initial image persist in the final generated image. This property can be exploited to employ Stable Diffusion as an image upscaler, thereby enabling image super-resolution. Let's explore this process step by step.

We will begin by introducing the concept of one-step super-resolution, followed by an exploration of multiple-step super-resolution.

One-step super-resolution

In this section, we will cover a solution to upscale images using image-to-image diffusion once. Here are the step-by-step instructions to implement it:

1. Let's start by generating a 256x256 start image using Stable Diffusion. Instead of downloading an image from the internet or utilizing an external image as the input, let's leverage Stable Diffusion to generate one. After all, this is the area where Stable Diffusion excels:

   ```
   import torch
   from diffusers import StableDiffusionPipeline
   ```

```
text2img_pipe = StableDiffusionPipeline.from_pretrained(
    "stablediffusionapi/deliberate-v2",
    torch_dtype = torch.float16
).to("cuda:0")

prompt = "a realistic photo of beautiful woman face"
neg_prompt = "NSFW, bad anatomy"

raw_image = text2img_pipe(
    prompt = prompt,
    negative_prompt = neg_prompt,
    height = 256,
    width = 256,
    generator = torch.Generator("cuda").manual_seed(3)
).images[0]
display(raw_image)
```

The preceding code will generate an image, as shown in *Figure 11.1*:

Figure 11.1: A photo of a woman's face sized 256x256, generated by Stable Diffusion

You may not be able to see the noise and blur in the image if you view it in a printed form (e.g., a paper book). However, if you run the preceding code and enlarge the image, you can easily notice the *blur* and *noise* in the generated image.

Let's save the image for further processing:

```
image_name = "woman_face"
file_name_256x256 =f"input_images/{image_name}_256x256.png"
raw_image.save(file_name_256x256)
```

2. Resize the image to the target size. Initially, we need to establish a function for image adjustment, ensuring that the image's width and height are both divisible by 8:

```python
def get_width_height(width, height):
    width = (width//8)*8
    height = (height//8)*8
    return width,height
```

Next, resize it to the target size using image interpolation:

```python
from diffusers.utils import load_image
from PIL import Image

def resize_img(img_path,upscale_times):
    img              = load_image(img_path)
    if upscale_times <=0:
        return img
    width,height = img.size
    width = width * upscale_times
    height = height * upscale_times
    width,height = get_width_height(int(width),int(height))
    img = img.resize(
        (width,height),
        resample = Image.LANCZOS if upscale_times > 1 \
            else Image.AREA
    )
    return img
```

The **PIL (Pillow** [12]) `resize` image function scales the pixels to the desired dimensions. In the preceding code snippet, we utilize the `Image.LANCZOS` interpolation method.

The following code employs the `resize_img` function to resize an image threefold:

```python
resized_raw_image = resize_img(file_name_256x256, 3.0)
```

Feel free to input any floating-point number greater than `1.0` into the function.

3. Create an img-to-img pipeline as the upscaler. To enable guided image super-resolution, we need to provide a guided prompt, as shown here:

```python
sr_prompt = """8k, best quality, masterpiece, realistic, photo-
realistic, ultra detailed, sharp focus, raw photo, """
prompt = """
a realistic photo of beautiful woman face
"""
prompt = f"{sr_prompt}{prompt}"
neg_prompt = "worst quality, low quality, lowres, bad anatomy"
```

sr_prompt means **super-resolution prompt** and can be resused without changing the same for any super-resolution tasks. Next, call the pipeline to upscale the image:

```
prompt = f"{sr_prompt}{prompt}"

neg_prompt = "worst quality, low quality, lowres, bad anatomy"

img2image_3x = img2img_pipe(
    image = resized_raw_image,
    prompt = prompt,
    negative_prompt = neg_prompt,
    strength = 0.3,
    num_inference_steps = 80,
    guidance_scale = 8,
    generator = torch.Generator("cuda").manual_seed(1)
).images[0]
img2image_3x
```

Note that the strength parameter is set to 0.3, meaning that each denoising step applies 30% Gaussian noise to the latent. When utilizing the plain text-to-image pipeline, the strength is set to 1.0 by default. By increasing the strength value here, more new elements will be introduced to the initial image. From my testing, 0.3 seems to strike a well-balanced point. However, you have the flexibility to adjust it to values such as 0.25 or elevate it to 0.4.

For an img-to-img pipeline from Diffusers, the actual denoising steps will be a multiplication of num_inference_steps by strength. The total denoising steps will be $80 \times 0.3 = 24$. This is not a rule enforced by Stable Diffusion; it is sourced from the implementation of Diffusers' Stable Diffusion pipeline.

As explained in *Chapter 3*, the guidance_scale parameter governs how closely the result aligns with the provided prompt and neg_prompt. In practice, a higher guidance_scale will yield a slightly clearer image but may also alter the image elements more, while a lower guidance_scale will lead to a more blurred image while preserving more original image elements. If you're uncertain about the value, opt for something between 7 and 8.

Once you run the preceding code, you'll observe that not only does the size of the original image upscale to 768x768 but the image quality also experiences a significant enhancement.

However, this is not the end; we can reuse the preceding process to further improve image resolution and quality.

Let's save the image for further usage:

```
file_name_768x768 = f"input_images/{image_name}_768x768.png"
img2image_3x.save(file_name_768x768)
```

Next, let's upscale the image using multiple image-to-image steps.

Multiple-step super-resolution

Using the one-step resolution, the code upscales the image from 256x256 to 768x768. In this section, we're taking the process a step further by increasing the image size to double its current dimensions.

Note that before progressing to an even higher resolution image, you need to ensure that the utilization of VRAM might necessitate more than 8 GB.

We will primarily be reusing the code from the one-step super-resolution process:

1. Double the size of the image:

```python
resized_raw_image = resize_img(file_name_768x768, 2.0)
display(resized_raw_image)
```

2. Further image super-resolution code can be applied to increase the resolution of an image by six times (256x256 to 1,536x1,536), which can significantly enhance the clarity and details of the image:

```python
sr_prompt = "8k, best quality, masterpiece, realistic, photo-
realistic, ultra detailed, sharp focus, raw photo,"

prompt = """
a realistic photo of beautiful woman face
"""
prompt = f"{sr_prompt}{prompt}"

neg_prompt = "worst quality, low quality, lowres, bad anatomy"

img2image_6x = img2img_pipe(
    image = resized_raw_image,
    prompt = prompt,
    negative_prompt = neg_prompt,
    strength = 0.3,
    num_inference_steps = 80,
    guidance_scale = 7.5,
    generator = torch.Generator("cuda").manual_seed(1)
).images[0]
img2image_6x
```

The preceding code will produce an image that is six times the size of the original, greatly enhancing its quality.

A super-resolution result comparison

Now, let's examine the six-times-upscaled image and compare it with the original image to see how much the image quality has improved.

Figure 11.2 provides a side-by-side comparison of the original image and the six-times-upscaled super-resolution image:

Figure 11.2: Left – the original raw image, and right – the image
with the six-times-upscaled super-resolution

Kindly check out the ebook version to easily discern the finer enhancements. *Figure 11.3* provides a clear depiction of the improvements in the mouth area:

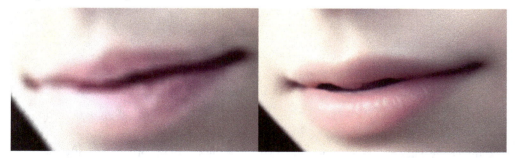

Figure 11.3: Left – the mouth from the original raw image, and right – the
mouth from the image with six-times-upscaled super-resolution

Figure 11.4 shows the improvements to the eyes:

Figure 11.4: Above – the eyes from the original raw image, and below – the
eyes from the image with six-times-upscaled super-resolution

Stable Diffusion enhances nearly every aspect of the image – from the eyebrows and eyelashes to the pupils – resulting in substantial improvements over the original raw image.

Img-to-Img limitations

The `deliberate-v2` Stable Diffusion model is a checkpoint model built on SD v1.5, and it's trained using 512x512 images. As a result, the img-to-img pipeline inherits all the constraints of this model. When attempting to upscale an image from 1,024x1,024 to an even higher resolution, the model might not be as efficient as it is with lower-resolution images.

However, img-to-img is not the only available solution to generate exceptionally high-quality images. Next, we will explore another technique that can upscale images with even greater detail.

ControlNet Tile image upscaling

Stable Diffusion ControlNet is a neural network architecture designed to enhance diffusion models through the incorporation of additional conditions. The concept behind this model stems from a paper titled *Adding Conditional Control to Text-to-Image Diffusion Models* [3], authored by Lvmin Zhang and Maneesh Agrawala in 2023. For more details about ControlNet, refer to *Chapter 13*.

ControlNet bears similarities to the image-to-image Stable Diffusion pipeline but boasts significantly greater potency.

When utilizing the img2img pipeline, we input the initial image, along with conditional text, to generate an image similar to the starting guidance image. In contrast, ControlNet employs one or more supplementary UNet models that work alongside the Stable Diffusion model. These UNet models process both the input prompt and the image concurrently, with results being merged back in each step of the UNet up-stage. A comprehensive exploration of ControlNet is provided in *Chapter 13*.

Compared with the image-to-image pipeline, ControlNet yields superior outcomes. Among the ControlNet models, ControlNet Tile stands out for its ability to upscale images by introducing substantial detail information to the original image.

In the subsequent code, we will employ the latest ControlNet version, 1.1. The authors of the paper and model affirm that they will maintain the architecture consistently up until ControlNet V1.5. At the time of reading, the most recent ControlNet iteration might surpass v1.1. There's a likelihood that you can repurpose the v1.1 code for later versions of ControlNet.

Steps to use ControlNet Tile to upscale an image

Next, let's use ControlNet Tile to upscale an image step by step:

1. Initialize ControlNet Tile model. The following code will start a ControlNet v1.1 model. Note that when ControlNet starts from v1.1, the sub-type of ControlNet is specified by `subfolder = 'control_v11f1e_sd15_tile'`:

    ```
    import torch
    from diffusers import ControlNetModel

    controlnet = ControlNetModel.from_pretrained(
        'takuma104/control_v11',
        subfolder = 'control_v11f1e_sd15_tile',
        torch_dtype = torch.float16
    )
    ```

    We can't simply use ControlNet itself to do anything; we will need to spin up a Stable Diffusion V1.5 pipeline to work together with the ControlNet model.

2. Initialize a Stable Diffusion v1.5 model pipeline. The primary advantage of using ControlNet lies in its compatibility with any checkpoint model that has been fine-tuned from the Stable Diffusion base model. We will continue to use the Stable Diffusion V1.5-based model due to its outstanding quality and relatively low VRAM requirements. Given these attributes, Stable Diffusion v1.5 is expected to retain its relevance for a considerable period:

    ```
    # load controlnet tile
    from diffusers import StableDiffusionControlNetImg2ImgPipeline

    # load checkpoint model with controlnet
    pipeline = StableDiffusionControlNetImg2ImgPipeline. \
    ```

```
from_pretrained(
    "stablediffusionapi/deliberate-v2",
    torch_dtype    = torch.float16,
    controlnet     = controlnet
)
```

In the provided code, we furnish `controlnet` which we initialized in *step 1* as a parameter for the `StableDiffusionControlNetImg2ImgPipeline` pipeline. Additionally, the code closely resembles that of a standard Stable Diffusion pipeline.

3. Resize the image. This is the same step we took in the image-to-image pipeline; we need to enlarge the image to the target size:

```
image_name = "woman_face"
file_name_256x256 = f"input_images/{image_name}_256x256.png"
resized_raw_image = resize_img(file_name_256x256, 3.0)
resized_raw_image
```

The preceding code upsizes the image three times using the LANCZOS interpolation:

```
Image super-resolution using ControlNet Tile
# upscale
sr_prompt = "8k, best quality, masterpiece, realistic, photo-
realistic, ultra detailed, sharp focus, raw photo,"

prompt = """
a realistic photo of beautiful woman face
"""

prompt = f"{sr_prompt}{prompt}"

neg_prompt = "worst quality, low quality, lowres, bad anatomy"

pipeline.to("cuda")
cn_tile_upscale_img = pipeline(
    image = resized_raw_image,
    control_image = resized_raw_image,
    prompt = prompt,
    negative_prompt = neg_prompt,
    strength = 0.8,
    guidance_scale = 7,
    generator = torch.Generator("cuda"),
    num_inference_steps = 50
).images[0]

cn_tile_upscale_img
```

We reuse the positive prompt and negative prompt from the image-to-image upscaler. The distinctions are outlined here:

- We assign the raw upscaled image to both the initial diffusion image, denoted as `image = resized_raw_image`, and the ControlNet start image, marked as `control_image = resized_raw_image`.

- The strength is configured to `0.8` in order to leverage the influence of ControlNet on denoising, thereby enhancing the generation process.

Note that we can lower the strength parameter to preserve as much original image as possible.

The ControlNet Tile upscaling result

With just a single round of three-fold super-resolution, we can significantly enhance our image by introducing an abundance of intricate details:

Figure 11.5: Left – the original raw image, and right – the ControlNet Tile three-times-upscaled super-resolution

When compared to the image-to-image upscaler, ControlNet Tile incorporates even more details. Upon zooming into the image, you can observe the addition of individual hair strands, leading to an overall improvement in image quality.

To achieve a comparable outcome, the image-to-image approach would require multiple steps to upscale the image sixfold. In contrast, ControlNet Tile accomplishes the same outcome with a single round of threefold upscaling.

Furthermore, ControlNet Tile offers the advantage of relatively lower VRAM usage compared to the image-to-image solution.

Additional ControlNet Tile upscaling samples

The ControlNet Tile super-resolution can produce remarkable results for a wide array of photos and images. Here are a few additional samples achieved by using just a few lines of code to generate, resize, and upscale images, capturing intricate details:

- **Man's face**: The code to generate, resize, and upscale this image is as follows:

```
# step 1. generate an image
prompt = """
Raw, analog a portrait of an 43 y.o. man ,
beautiful photo with highly detailed face by greg rutkowski and
magali villanueve
"""

neg_prompt = "NSFW, bad anatomy"

text2img_pipe.to("cuda")
raw_image = text2img_pipe(
    prompt = prompt,
    negative_prompt = neg_prompt,
    height = 256,
    width = 256,
    generator = torch.Generator("cuda").manual_seed(3)
).images[0]
display(raw_image)

image_name = "man"
file_name_256x256 = f"input_images/{image_name}_256x256.png"
raw_image.save(file_name_256x256)

# step 2. resize image
resized_raw_image = resize_img(file_name_256x256, 3.0)
display(resized_raw_image)

# step 3. upscale image
sr_prompt = "8k, best quality, masterpiece, realistic, photo-
realistic, ultra detailed, sharp focus, raw photo,"

prompt = f"{sr_prompt}{prompt}"

neg_prompt = "worst quality, low quality, lowres, bad anatomy"

pipeline.to("cuda")
cn_tile_upscale_img = pipeline(
```

```
    image = resized_raw_image,
    control_image = resized_raw_image,
    prompt = prompt,
    negative_prompt = neg_prompt,
    strength = 0.8,
    guidance_scale = 7,
    generator = torch.Generator("cuda"),
    num_inference_steps = 50,
    # controlnet_conditioning_scale = 0.8
).images[0]

display(cn_tile_upscale_img)
```

The result is shown in *Figure 11.6*:

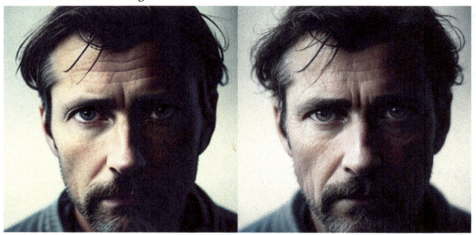

Figure 11.6: Left – the original raw image, and right – the ControlNet
Tile three-times-upscaled super-resolution

- **Old man**: Here is the code to generate, resize, and upscale the image:

```
# step 1. generate an image
prompt = """
A realistic photo of an old man, standing in the garden, flower
and green trees around, face view
"""

neg_prompt = "NSFW, bad anatomy"

text2img_pipe.to("cuda")
raw_image = text2img_pipe(
```

```
    prompt = prompt,
    negative_prompt = neg_prompt,
    height = 256,
    width = 256,
    generator = torch.Generator("cuda").manual_seed(3)
).images[0]
display(raw_image)

image_name = "man"
file_name_256x256 = f"input_images/{image_name}_256x256.png"
raw_image.save(file_name_256x256)

# step 2. resize image
resized_raw_image = resize_img(file_name_256x256, 4.0)
display(resized_raw_image)

# step 3. upscale image
sr_prompt = "8k, best quality, masterpiece, realistic, photo-
realistic, ultra detailed, sharp focus, raw photo,"

prompt = f"{sr_prompt}{prompt}"

neg_prompt = "worst quality, low quality, lowres, bad anatomy"

pipeline.to("cuda")
cn_tile_upscale_img = pipeline(
    image = resized_raw_image,
    control_image = resized_raw_image,
    prompt = prompt,
    negative_prompt = neg_prompt,
    strength = 0.8,
    guidance_scale = 7,
    generator = torch.Generator("cuda"),
    num_inference_steps = 50,
    # controlnet_conditioning_scale = 0.8
).images[0]

display(cn_tile_upscale_img)
```

The result is shown in *Figure 11.7*:

Figure 11.7: Left – the original raw image of the old man, and right – the
ControlNet Tile four-times-upscaled super-resolution

- **Royal female**: Here is the code to generate, resize, and upscale the image:

```
# step 1. generate an image
prompt = """
upper body photo of royal female, elegant, pretty face, majestic
dress,
sitting on a majestic chair, in a grand fantasy castle hall,
shallow depth of field, cinematic lighting, Nikon D850,
film still, HDR, 8k
"""

neg_prompt = "NSFW, bad anatomy"

text2img_pipe.to("cuda")
raw_image = text2img_pipe(
    prompt = prompt,
    negative_prompt = neg_prompt,
    height = 256,
    width = 256,
    generator = torch.Generator("cuda").manual_seed(7)
).images[0]
display(raw_image)

image_name = "man"
file_name_256x256 = f"input_images/{image_name}_256x256.png"
raw_image.save(file_name_256x256)
```

```
# step 2. resize image
resized_raw_image = resize_img(file_name_256x256, 4.0)
display(resized_raw_image)

# step 3. upscale image
sr_prompt = "8k, best quality, masterpiece, realistic, photo-
realistic, ultra detailed, sharp focus, raw photo,"

prompt = f"{sr_prompt}{prompt}"

neg_prompt = "worst quality, low quality, lowres, bad anatomy"

pipeline.to("cuda")
cn_tile_upscale_img = pipeline(
    image = resized_raw_image,
    control_image = resized_raw_image,
    prompt = prompt,
    negative_prompt = neg_prompt,
    strength = 0.8,
    guidance_scale = 7,
    generator = torch.Generator("cuda"),
    num_inference_steps = 50,
    # controlnet_conditioning_scale = 0.8
).images[0]

display(cn_tile_upscale_img)
```

The result is shown in *Figure 11.8*:

Figure 11.8: Left – the original raw royal female, and right – the
ControlNet Tile four- times-upscaled super-resolution

Summary

This chapter offered a summary of contemporary image upscaling and super-resolution methodologies, emphasizing their unique characteristics. The primary focus of the chapter was on two super-resolution techniques that leverage the capabilities of Stable Diffusion:

- Utilizing the Stable Diffusion image-to-image pipeline
- Implementing ControlNet Tile to upscale images while enhancing details

Furthermore, we showcased additional examples of the ControlNet Tile super-resolution technique.

If your goal is to preserve as many aspects of the original image as possible during upscaling, we recommend the image-to-image pipeline. Conversely, if you prefer an AI-driven approach that generates a wealth of detail, ControlNet Tile is the more appropriate option.

In *Chapter 12*, we will develop a scheduled prompt parser to allow us more accurate control over image generation.

References

1. Hugging Face – Super-Resolution: `https://huggingface.co/docs/diffusers/v0.13.0/en/api/pipelines/stable_diffusion/upscale`

2. Hugging Face – Ultra-fast ControlNet with Diffusers: `https://huggingface.co/blog/controlnet`

3. Lvmin Zhang, Maneesh Agrawala, Adding Conditional Control to Text-to-Image Diffusion Models: `https://arxiv.org/abs/2302.05543`

4. Lvmin Zhang, ControlNet original implementation code: `https://github.com/lllyasviel`

5. Lvmin Zhang, ControlNet 1.1 Tile: `https://github.com/lllyasviel/ControlNet-v1-1-nightly#controlnet-11-tile`

6. Nearest-neighbor interpolation: `https://en.wikipedia.org/wiki/Nearest-neighbor_interpolation`

7. Bilinear interpolation: `https://en.wikipedia.org/wiki/Bilinear_interpolation`

8. Bicubic interpolation: `https://en.wikipedia.org/wiki/Bicubic_interpolation`

9. Lanczos resampling: `https://en.wikipedia.org/wiki/Lanczos_resampling`

10. ESRGAN: Enhanced Super-Resolution Generative Adversarial Networks: `https://arxiv.org/abs/1809.00219`

11. SwinIR: Image Restoration Using Swin Transformer: `https://arxiv.org/abs/2108.10257`

12. Python Pillow package: `https://github.com/python-pillow/Pillow`

12

Scheduled Prompt Parsing

In *Chapter 10*, we discussed how to unlock the 77-token prompt limitation and a solution to enable prompt weighting, which paved the way for this chapter. With the knowledge from *Chapter 10*, we can generate various kinds of images by leveraging the power of natural language and weighting formats. However, there are some limitations inherent in the out-of-the-box code from the Hugging Face Diffusers package.

For example, we cannot write a prompt to ask Stable Diffusion to generate a cat in the first five steps and then a dog in the next five steps. Similarly, we cannot write a prompt to ask Stable Diffusion to blend two concepts by alternately denoising the two concepts.

In this chapter, we will explore the two solutions in the following topics:

* Using the Compel package
* Building a custom scheduled prompt pipeline

Technical requirements

To get started with the code in this chapter, you will need to install the necessary packages for running Stable Diffusion. For detailed instructions on how to set up these packages, refer to *Chapter 2*.

In addition to the packages required by Stable Diffusion, you will also need to install the `Compel` package for the *Using the Compel package* section, and the `lark` package for the *Building a custom scheduled prompt pipeline* section.

I will provide step-by-step instructions for installing and using these packages in each section.

Using the Compel package

Compel [1] is an open source text prompt-weighting and blending library developed and maintained by Damian Stewart. It is one of the easiest ways to enable blending prompts in Diffusers. This package also has the capability to apply weighting to prompts, similar to the solution we implemented in

Chapter 10, but with a different weighting syntax. In this chapter, I will introduce the blending feature that can help us write a prompt to generate an image with two or more concepts blended.

Imagine that we want to create a photo that is half cat and half dog. How would we do it with prompts? Let's say we simply give Stable Diffusion the following prompt:

```
A photo with half cat and half dog
```

Here are the lines of Python code (without using Compel):

```python
import torch
from diffusers import StableDiffusionPipeline
pipeline = StableDiffusionPipeline.from_pretrained(
    "stablediffusionapi/deliberate-v2",
    torch_dtype = torch.float16,
    safety_checker = None
).to("cuda:0")
image = pipeline(
    prompt = "A photo with half cat and half dog",
    generator = torch.Generator("cuda:0").manual_seed(3)
).images[0]
image
```

You will see the result shown in *Figure 12.1*:

Figure 12.1: Result of the A photo with half cat and half dog prompt

The word `half` should be applied to the photo itself, rather than the image. In this case, we can use Compel to help generate a text embedding that blends cat and dog.

Before importing the `Compel` package, we will need to install the package:

```
pip install compel
```

Note that the reason the `Compel` package works with Diffusers is that the package produces text embedding using `tokenizer` (type: `transformers.models.clip.tokenization_clip.CLIPTokenizer`) and `text_encoder` (type: `transformers.models.clip.modeling_clip.CLIPTextModel`) from the Stable Diffusion model file. We should also be aware of this during the initialization of the `Compel` object:

```
from comp
compel = Compel(
    tokenizer = pipeline.tokenizer,
    text_encoder = pipeline.text_encoder
)
```

The `pipeline` (type: `StableDiffusionPipeline`) is the Stable Diffusion pipeline we just created. Next, create a blend prompt using the following format:

```
prompt = '("A photo of cat", "A photo of dog").blend(0.5, 0.5)'
prompt_embeds = compel(prompt)
```

Then, feed the text embedding into the Stable Diffusion pipeline through the `prompt_embeds` parameter:

```
image = pipeline(
    prompt_embeds = prompt_embeds,
    generator = torch.Generator("cuda:0").manual_seed(1)
).images[0]
image
```

We will see a pet that looks like a cat and also a dog, as shown in *Figure 12.2*:

Figure 12.2: A blended photo – half cat and half dog – using Compel

We can change the proposition number of the blend to have more `cat` or more `dog`. Let's change the prompt to this:

```
prompt = '("A photo of cat", "A photo of dog").blend(0.7, 0.3)'
```

We will get a photo slightly more like a cat, as shown in *Figure 12.3*:

Figure 12.3: A blended photo of a cat and dog using Compel – 70% cat, 30% dog

Compel can do more than just prompt blending; it can also provide `weighted` and `and` conjunction prompts. You can explore more usage examples and features in its Syntax Features [2] documentation.

While it is easy to use Compel to blend prompts, as we have seen in the previous example, a blending prompt like the one that follows is strange and unintuitive for day-to-day use:

```
prompt = '("A photo of cat", "A photo of dog").blend(0.7, 0.3)'
```

Upon my initial review of the sample code from the Compel repository, I was intrigued by the following line: `("A photo of cat", "A photo of dog").blend(0.7, 0.3)`. This string prompts several questions, such as how can the `blend()` function be invoked? However, it becomes clear that `blend()` is part of the prompt string and not a function that can be invoked within the Python code.

In contrast, the prompt blending or scheduling feature of Stable Diffusion WebUI [3] is relatively more user-friendly. The syntax allows us to achieve the same blending effect with a prompt syntax like this:

```
[A photo of cat:A photo of dog:0.5]
```

This scheduled prompt in Stable Diffusion WebUI will render a photo of a cat during the first 50% of the steps and a photo of a dog during the last 50% of the steps.

Alternatively, you can use the | operator to alternate the prompt:

```
[A photo of cat|A photo of dog]
```

The preceding scheduled prompt will alternate between rendering photos of a cat and a dog. In other words, it will render a photo of a cat in one step and a photo of a dog in the next step, continuing this pattern until the end of the entire rendering process.

These two scheduling features can also be achieved by Diffusers. In the following section, we will explore how to implement both of these advanced prompt scheduling features for Diffusers.

Building a custom scheduled prompt pipeline

As we discussed in *Chapter 5*, the generation process utilizes input prompt embedding to denoise an image at each step. By default, every denoising step employs the exact same embedding. However, to gain more precise control over the generation, we can modify the pipeline code to supply unique embeddings for each denoising step.

Take, for instance, the following prompt:

```
[A photo of cat:A photo of dog:0.5]
```

During a total of 10 denoising steps, we hope the pipeline can remove noise in the first 5 steps to reveal A photo of cat, and the following 5 steps to reveal A photo of dog. To make this happen, we will need to implement the following components:

- A prompt parser capable of extracting the scheduling number from the prompt

- A method to embed the prompts and create a list of prompt embeddings that matches the number of steps

- A new pipeline class derived from the Diffusers pipeline, enabling us to incorporate our new functionality into the pipeline while preserving all existing features of the Diffusers pipeline

Next, let's implement the formatted prompt parser.

A scheduled prompt parser

The open sourced Stable Diffusion WebUI project's source code reveals that we can use lark [4] – a parsing toolkit for Python. We will also use the lark package to parse the scheduled prompt for our own prompt parser.

To install lark, run the following command:

```
pip install -U lark
```

The Stable Diffusion WebUI compatible prompt is defined in the following code:

```
import lark
schedule_parser = lark.Lark(r"""
!start: (prompt | /[][() :]/+)*
prompt: (emphasized | scheduled | alternate | plain | WHITESPACE)*
!emphasized: "(" prompt ")"
        | "(" prompt ":" prompt ")"
        | "[" prompt "]"
scheduled: "[" [prompt ":"] prompt ":" [WHITESPACE] NUMBER "]"
alternate: "[" prompt ("|" prompt)+ "]"
WHITESPACE: /\s+/
plain: /([^\\\[\]() :|]|\\.)+/
%import common.SIGNED_NUMBER -> NUMBER
""")
```

If you decide to get up to your neck in the syntax swamp to fully understand every line of the definition, the `lark` document [5] is the place to go.

Next, we'll use the Python function from the SD WebUI code repository. This function utilizes the Lark `schedule_parser` syntax definition to parse the input prompt:

```
def get_learned_conditioning_prompt_schedules(prompts, steps):
    def collect_steps(steps, tree):
        l = [steps]
        class CollectSteps(lark.Visitor):
            def scheduled(self, tree):
                tree.children[-1] = float(tree.children[-1])
                if tree.children[-1] < 1:
                    tree.children[-1] *= steps
                tree.children[-1] = min(steps, int(tree.children[-1]))
                l.append(tree.children[-1])
            def alternate(self, tree):
                l.extend(range(1, steps+1))
        CollectSteps().visit(tree)
        return sorted(set(l))

    def at_step(step, tree):
        class AtStep(lark.Transformer):
            def scheduled(self, args):
                before, after, _, when = args
                yield before or () if step <= when else after
            def alternate(self, args):
                yield next(args[(step - 1)%len(args)])
```

```
        def start(self, args):
            def flatten(x):
                if type(x) == str:
                    yield x
                else:
                    for gen in x:
                        yield from flatten(gen)
            return ''.join(flatten(args))
        def plain(self, args):
            yield args[0].value
        def __default__(self, data, children, meta):
            for child in children:
                yield child
    return AtStep().transform(tree)

def get_schedule(prompt):
    try:
        tree = schedule_parser.parse(prompt)
    except lark.exceptions.LarkError as e:
        if 0:
            import traceback
            traceback.print_exc()
        return [[steps, prompt]]
    return [[t, at_step(t, tree)] for t in collect_steps(steps,
        tree)]

promptdict = {prompt: get_schedule(prompt) for prompt in
    set(prompts)}
return [promptdict[prompt] for prompt in prompts]
```

Set the total denoising steps to 10, and give a shorter name, g, for this function:

```
steps = 10
g = lambda p: get_learned_conditioning_prompt_schedules([p], steps)[0]
```

Now, let's throw some prompts to the function to see the parsing results:

- Test #1: cat:

  ```
  g("cat")
  ```

  The preceding code will parse the cat input text as the following string:

  ```
  [[10, 'cat']]
  ```

  The result indicates that all 10 steps will use cat to generate the image.

- Test #2: `[cat:dog:0.5]`:

  Change the prompt to `[cat:dog:0.5]`:

  ```
  g('[cat:dog:0.5]')
  ```

  The function will generate the following result:

  ```
  [[5, 'cat'], [10, 'dog']]
  ```

  The result means using cat for the first 5 steps, and dog for the last 5 steps.

- Test #3: `[cat|dog]`:

  The function also supports alternative scheduling. Change the prompt to `[cat  |  dog]`, with an "or" | operator in the middle of the two names:

  ```
  g('[cat|dog]')
  ```

  The prompt parser will generate the following result:

  ```
  [[1, 'cat'],
   [2, 'dog'],
   [3, 'cat'],
   [4, 'dog'],
   [5, 'cat'],
   [6, 'dog'],
   [7, 'cat'],
   [8, 'dog'],
   [9, 'cat'],
   [10, 'dog']]
  ```

In other words, it alternates the prompt for each denoising step.

So far, it works well in terms of parsing. However, before feeding it to the pipeline, additional work needs to be done.

Filling in the missing steps

In *Test #2*, the generated result includes only two elements. We need to expand the result list to cover each step:

```
def parse_scheduled_prompts(text, steps=10):
    text = text.strip()
    parse_result = None
    try:
        parse_result = get_learned_conditioning_prompt_schedules(
            [text],
            steps = steps
        )[0]
```

```
        except Exception as e:
            print(e)

    if len(parse_result) == 1:
        return parse_result

    prompts_list = []

    for i in range(steps):
        current_prompt_step, current_prompt_content = \
            parse_result[0][0],parse_result[0][1]
        step = i + 1
        if step < current_prompt_step:
            prompts_list.append(current_prompt_content)
            continue

        if step == current_prompt_step:
            prompts_list.append(current_prompt_content)
            parse_result.pop(0)

    return prompts_list
```

This Python function, `parse_scheduled_prompts`, takes two arguments: `text` and `steps` (with a default value of 10). The function processes the given text to generate a list of prompts based on a learned conditioning schedule.

Here's a step-by-step explanation of what the function does:

1. Use a `try-except` block to call the `get_learned_conditioning_prompt_schedules` function with the processed text and the specified number of steps. The result is stored in `parse_result`. If there's an exception – say, a syntax error, it will be caught and printed.

2. If the length of `parse_result` is 1, return `parse_result` as the final output.

3. Loop through the range of steps and perform the following actions:

   I.   Get the current prompt step and content from `parse_result`.

   II.   Increment the loop counter `i` by 1 and store it in the variable step.

   III. If `step` is less than the current prompt step, append the current prompt content to `prompts_list` and continue to the next iteration.

   IV.   If `step` is equal to the current prompt step, append the current prompt content to `prompts_list` and remove the first element from `parse_result`.

4. Return the `prompts_list` as the final output.

The function essentially generates a list of prompts based on the learned conditioning schedule, with each prompt being added to the list according to the specified number of steps.

Let's call this function to test it out:

```
prompt_list = parse_scheduled_prompts("[cat:dog:0.5]")
prompt_list
```

We will get a prompt list as shown here:

```
['cat',
 'cat',
 'cat',
 'cat',
 'cat',
 'dog',
 'dog',
 'dog',
 'dog',
 'dog']
```

Five prompts for `cat`, and five prompts for `dog` – each denoising step will take one of the prompts.

A Stable Diffusion pipeline supporting scheduled prompts

Until now, all prompts are still in plain text form. We will need to use custom embedding code to encode unlimited and weighted prompts into embeddings, or we can simply use the default encoder from Diffusers to generate embeddings but with a 77-token limitation.

To make the logic easier and more concise to follow, we will use the default text encoder in this section. Once we figure out how it works, it will be easy to swap the encoder with the more powerful one we built in *Chapter 10*.

Since we will perform a minor surgical operation on the original Diffusers Stable Diffusion pipeline to support this embedding list, the operation includes creating a new pipeline class inherited from the Diffusers pipeline. We can directly reuse the tokenizer and text encoder from the initialized pipeline by using the following code:

```
...
prompt_embeds = self._encode_prompt(
    prompt,
    device,
    num_images_per_prompt,
    do_classifier_free_guidance,
    negative_prompt,
```

```
        negative_prompt_embeds=negative_prompt_embeds,
    )
...
```

I will further explain the preceding code next. We will implement the whole logic in the `scheduler_call` function (similar to the `__call__` function of `StableDiffusionPipeline`):

```python
from typing import List, Callable, Dict, Any
from torch import Generator, FloatTensor
from diffusers.pipelines.stable_diffusion import (
    StableDiffusionPipelineOutput)
from diffusers import (
    StableDiffusionPipeline, EulerDiscreteScheduler)

class StableDiffusionPipeline_EXT(StableDiffusionPipeline):
    @torch.no_grad()
    def scheduler_call(
        self,
        prompt: str | List[str] = None,
        height: int | None = 512,
        width: int | None = 512,
        num_inference_steps: int = 50,
        guidance_scale: float = 7.5,
        negative_prompt: str | List[str] | None = None,
        num_images_per_prompt: int | None = 1,
        eta: float = 0,
        generator: Generator | List[Generator] | None = None,
        latents: FloatTensor | None = None,
        prompt_embeds: FloatTensor | None = None,
        negative_prompt_embeds: FloatTensor | None = None,
        output_type: str | None = "pil",
        callback: Callable[[int, int, FloatTensor], None] | None = None,
        callback_steps: int = 1,
        cross_attention_kwargs: Dict[str, Any] | None = None,
    ):
        ...


        # 6. Prepare extra step kwargs. TODO: Logic should ideally
        # just be moved out of the pipeline
        extra_step_kwargs = self.prepare_extra_step_kwargs(
            generator, eta)

        # 7. Denoising loop
```

```python
num_warmup_steps = len(timesteps) - num_inference_steps * \
    self.scheduler.order
with self.progress_bar(total=num_inference_steps) as \
    progress_bar:
    for i, t in enumerate(timesteps):
        # AZ code to enable Prompt Scheduling,
        # will only function when
        # when there is a prompt_embeds_l provided.
        prompt_embeds_l_len = len(embedding_list)
        if prompt_embeds_l_len > 0:
            # ensure no None prompt will be used
            pe_index = (i)%prompt_embeds_l_len
            prompt_embeds = embedding_list[pe_index]

        # expand the latents if we are doing classifier
        #free guidance
        latent_model_input = torch.cat([latents] * 2) \
            if do_classifier_free_guidance else latents
        latent_model_input = self.scheduler. \
            scale_model_input(latent_model_input, t)

        # predict the noise residual
        noise_pred = self.unet(
            latent_model_input,
            t,
            encoder_hidden_states=prompt_embeds,
            cross_attention_kwargs=cross_attention_kwargs,
        ).sample

        # perform guidance
        if do_classifier_free_guidance:
            noise_pred_uncond, noise_pred_text = \
                noise_pred.chunk(2)
            noise_pred = noise_pred_uncond + guidance_scale * \
                (noise_pred_text - noise_pred_uncond)

        # compute the previous noisy sample x_t -> x_t-1
        latents = self.scheduler.step(noise_pred, t, latents,
            **extra_step_kwargs).prev_sample

        # call the callback, if provided
        if i == len(timesteps) - 1 or ((i + 1) > \
        num_warmup_steps and (i + 1) % \
        self.scheduler.order == 0):
```

```
                        progress_bar.update()
                        if callback is not None and i % callback_steps== 0:
                            callback(i, t, latents)

        if output_type == "latent":
            image = latents
        elif output_type == "pil":
            # 8. Post-processing
            image = self.decode_latents(latents)
            image = self.numpy_to_pil(image)
        else:
            # 8. Post-processing
            image = self.decode_latents(latents)

        if hasattr(self, "final_offload_hook") and \
            self.final_offload_hook is not None:
            self.final_offload_hook.offload()

        return StableDiffusionPipelineOutput(images=image)
```

This Python function, `scheduler_call`, is a method of the `StableDiffusionPipeline_EXT` class, which is a subclass of `StableDiffusionPipeline`.

Here are the steps to implement the whole logic:

1. Set the default scheduler to `EulerDiscreteScheduler` for a better generation result:

    ```
    if self.scheduler._class_name == "PNDMScheduler":
        self.scheduler = EulerDiscreteScheduler.from_config(
            self.scheduler.config
        )
    ```

2. Prepare the `device` and `do_classifier_free_guidance` parameters:

    ```
    device = self._execution_device
    do_classifier_free_guidance = guidance_scale > 1.0
    ```

3. Call the `parse_scheduled_prompts` function to have the `prompt_list` prompt list. This is the function we built in the previous section of this chapter:

    ```
    prompt_list = parse_scheduled_prompts(prompt)
    ```

4. If no scheduled prompt is found, use the normal single-prompt logic:

    ```
    embedding_list = []
    if len(prompt_list) == 1:
        prompt_embeds = self._encode_prompt(
    ```

```
            prompt,
            device,
            num_images_per_prompt,
            do_classifier_free_guidance,
            negative_prompt,
            negative_prompt_embeds=negative_prompt_embeds,
        )
    else:
        for prompt in prompt_list:
            prompt_embeds = self._encode_prompt(
                prompt,
                device,
                num_images_per_prompt,
                do_classifier_free_guidance,
                negative_prompt,
                negative_prompt_embeds=negative_prompt_embeds,
            )
            embedding_list.append(prompt_embeds)
```

In *step 4*, the function processes the input prompt(s) to generate the prompt embeddings. The input prompt can be a single string or a list of strings. The function first parses the input prompt(s) into a list called `prompt_list`. If there is only one prompt in the list, the function directly encodes the prompt using the `_encode_prompt` method and stores the result in `prompt_embeds`. If there are multiple prompts, the function iterates through `prompt_list` and encodes each prompt separately using the `_encode_prompt` method. The resulting embeddings are stored in `embedding_list`.

5. Prepare timesteps for the diffusion process:

```
self.scheduler.set_timesteps(num_inference_steps, device=device)
timesteps = self.scheduler.timesteps
```

6. Prepare latent variables to initialize the `latents` tensor (this is a PyTorch tensor):

```
num_channels_latents = self.unet.in_channels
batch_size = 1
latents = self.prepare_latents(
    batch_size * num_images_per_prompt,
    num_channels_latents,
    height,
    width,
    prompt_embeds.dtype,
    device,
    generator,
    latents,
)
```

7.   Implement the denoising loop:

```
num_warmup_steps = len(timesteps) - num_inference_steps * \
    self.scheduler.order
with self.progress_bar(total=num_inference_steps) as \
    progress_bar:
    for i, t in enumerate(timesteps):
        # custom code to enable Prompt Scheduling,
        # will only function when
        # when there is a prompt_embeds_l provided.
        prompt_embeds_l_len = len(embedding_list)
        if prompt_embeds_l_len > 0:
            # ensure no None prompt will be used
            pe_index = (i)%prompt_embeds_l_len
            prompt_embeds = embedding_list[pe_index]

        # expand the latents if we are doing
        # classifier free guidance
        latent_model_input = torch.cat([latents] * 2)
            if do_classifier_free_guidance else latents
        latent_model_input =
            self.scheduler.scale_model_input(
                latent_model_input, t)

        # predict the noise residual
        noise_pred = self.unet(
            latent_model_input,
            t,
            encoder_hidden_states=prompt_embeds,
            cross_attention_kwargs=cross_attention_kwargs,
        ).sample

        # perform guidance
        if do_classifier_free_guidance:
            noise_pred_uncond, noise_pred_text = \
                noise_pred.chunk(2)
            noise_pred = noise_pred_uncond + guidance_scale * \
                (noise_pred_text - noise_pred_uncond)

        # compute the previous noisy sample x_t -> x_t-1
        latents = self.scheduler.step(noise_pred, t,
            latents).prev_sample

        # call the callback, if provided
```

```
    if i == len(timesteps) - 1 or ((i + 1) >
        num_warmup_steps and (i + 1) %
        self.scheduler.order == 0):
        progress_bar.update()
        if callback is not None and i % callback_steps == 0:
            callback(i, t, latents)
```

In *step 7*, the denoising loop iterates through the timesteps of the diffusion process. If prompt scheduling is enabled (i.e., there are multiple prompt embeddings in embedding_list), the function selects the appropriate prompt embedding for the current timestep. The length of embedding_list is stored in prompt_embeds_l_len. If prompt_embeds_l_len is greater than 0, it means prompt scheduling is enabled. The function calculates the pe_index index for the current timestep, i, using the modulo operator (%). This ensures that the index wraps around the length of embedding_list and selects the appropriate prompt embedding for the current timestep. The selected prompt embedding is then assigned to prompt_embeds.

8. The last step is denoising post-processing:

```
image = self.decode_latents(latents)
image = self.numpy_to_pil(image)
return StableDiffusionPipelineOutput(images=image,
    nsfw_content_detected=None)
```

In the last step, we convert the image data from latent space to pixel space by calling the decode_latents() function. The StableDiffusionPipelineOutput class is used here for a consistent structure to be maintained when returning outputs from the pipeline. We use it here to make our pipeline compatible with the Diffusers pipeline. You can also find the complete code in the code files associated with this chapter.

Congratulations to you if you are still here! Let's execute it to witness the result:

```
pipeline = StableDiffusionPipeline_EXT.from_pretrained(
    "stablediffusionapi/deliberate-v2",
    torch_dtype = torch.float16,
    safety_checker = None
).to("cuda:0")
prompt = "high quality, 4k, details, A realistic photo of cute \
[cat:dog:0.6]"
neg_prompt = "paint, oil paint, animation, blur, low quality, \
bad glasses"
image = pipeline.scheduler_call(
    prompt = prompt,
    negative_prompt = neg_prompt,
    generator = torch.Generator("cuda").manual_seed(1)
).images[0]
image
```

We should see an image like the one shown in *Figure 12.4*:

Figure 12.4: A blended photo with 60% cat and 40% dog, using a custom scheduled prompt pipeline

Here's another example, using an alternative prompt [cat|dog]:

```
prompt = "high quality, 4k, details, A realistic photo of white \
[cat|dog]"
neg_prompt = "paint, oil paint, animation, blur, low quality, bad \
glasses"
image = pipeline.scheduler_call(
    prompt = prompt,
    negative_prompt = neg_prompt,
    generator = torch.Generator("cuda").manual_seed(3)
).images[0]
image
```

Our alternative prompt gives an image similar to *Figure 12.5*:

Figure 12.5: A photo of a blended cat and dog, using the alternative prompt
scheduling from our custom-scheduled prompt pipeline

If you see half-cat, half-dog images generated, as shown in *Figure 12.4* and *Figure 12.5*, you have successfully built your custom prompt scheduler.

Summary

In this chapter, we introduced two solutions for conducting scheduled prompt image generation. The first solution, the `Compel` package, is the easiest one to use. Simply install the package, and you can use its prompt blend feature to blend two or more concepts in one prompt.

The second solution is a customized pipeline that first parses the prompt string and prepares a prompt list for each denoising step. The custom pipeline loops through the prompt list to create an embedding list. Finally, a `scheduler_call` function uses the prompt embedding from the embedding list to generate images with precise control.

If you successfully implement the custom scheduled pipeline, you can control generation in a more precise way. Speaking of controlling, in *Chapter 13*, we are going to explore another way to control image generation – ControlNet.

References

1. Compel: `https://github.com/damian0815/compel`

2. Compel Syntax Features: `https://github.com/damian0815/compel/blob/main/doc/syntax.md`

3. Stable Diffusion WebUI Prompt Editing: `https://github.com/AUTOMATIC1111/stable-diffusion-webui/wiki/Features#prompt-editing`

4. Lark – a parsing toolkit for Python: `https://github.com/lark-parser/lark`

5. Lark usage document: `https://lark-parser.readthedocs.io/en/stable/`

Part 3 – Advanced Topics

In Parts 1 and 2, we established a solid foundation for Stable Diffusion, covering its fundamentals, customization options, and optimization techniques. Now, it's time to venture into more advanced territories, where we'll explore cutting-edge applications, innovative models, and expert-level strategies to generate remarkable visual content.

The chapters in this part will take you on a thrilling journey through the latest developments in Stable Diffusion. You'll learn how to generate images with unprecedented control using ControlNet, craft captivating videos with AnimateDiff, and extract insightful descriptions from images using powerful vision-language models such as BLIP-2 and LLaVA. Additionally, you'll get acquainted with Stable Diffusion XL, a newer and more capable iteration of the Stable Diffusion model.

To top it off, we'll delve into the art of crafting optimized prompts for Stable Diffusion, including techniques to write effective prompts and leverage large language models to automate the process. By the end of this part, you'll possess the expertise to tackle complex projects, push the boundaries of Stable Diffusion, and unlock new creative possibilities. Get ready to unleash your full potential and produce breathtaking results!

This part contains the following chapters:

- *Chapter 13, Generating Images with ControlNet*
- *Chapter 14, Generating Video Using Stable Diffusion*
- *Chapter 15, Generating Image Descriptions Using BLIP-2 and LLaVA*
- *Chapter 16, Exploring Stable Diffusion XL*
- *Chapter 17, Building Optimized Prompts for Stable Diffusion*

13

Generating Images with ControlNet

Stable Diffusion's ControlNet is a neural network plugin that allows you to control diffusion models by adding extra conditions. It was first introduced in a paper called Adding Conditional Control to Text-to-Image Diffusion Models [1] by Lvmin Zhang and Maneesh Agrawala, published in 2023.

This chapter will cover the following topics:

- What is ControlNet and how is it different?
- Usage of ControlNet
- Using multiple ControlNets in one pipeline
- How ControlNet works
- More ControlNet usage

By the end of this chapter, you will understand how ControlNet works and how to use Stable Diffusion V1.5 and Stable Diffusion XL ControlNet models.

What is ControlNet and how is it different?

In terms of "control," you may recall textual embedding, LoRA, and the image-to-image diffusion pipeline. But what makes ControlNet different and useful?

Unlike other solutions, ControlNet is a model that works on the UNet diffusion process directly. We compare these solutions in *Table 13.1*:

Control Method	Functioning Stage	Usage Scenario
Textual Embedding	Text encoder	Add a new style, a new concept, or a new face
LoRA	Merge LoRA weights to the UNet model (and the CLIP text encoder, optional)	Add a set of styles, concepts, and generate content
Image-to-Image	Provide the initial latent image	Fix images, or add styles and concepts to images
ControlNet	ControlNet participant denoising together with a checkpoint model UNet	Control shape, pose, content detail

Table 13.1: A comparison of textual embedding, LoRA, image-to-image, and ControlNet

In many ways, ControlNet is similar to the image-to-image pipeline, as we discussed in *Chapter 11*. Both image-to-image and ControlNet can be used to enhance images.

However, ControlNet can "control" the image in a more precise way. Imagine you want to generate an image that uses a specific pose from another image or perfectly align objects within the scene to a specific reference point. This kind of precision is impossible with the out-of-the-box Stable Diffusion model. ControlNet is the tool that can help you achieve these goals.

Besides, ControlNet models work with all other open source checkpoint models, unlike some other solutions, which work only with one base model provided by their author. The team that created ControlNet not only open-sourced the model but also open-sourced the code to train a new model. In other words, we can train a ControlNet model and make it work with any other model. This is what the original paper says [1]:

> *Since Stable Diffusion is a typical UNet structure, this ControlNet architecture is likely to be applicable with other models.*

Note that ControlNet models will only work with models using the same base model. A **Stable Diffusion (SD) v1.5** ControlNet model works with all other SD v1.5 models. For **Stable Diffusion XL (SDXL)** models, we will need a ControlNet model that is trained with SDXL. This is because SDXL models use a different architecture, a larger UNet than the SD v1.5. Without additional work, a ControlNet model is trained with one architecture and only works with this type of model.

I used "*without additional work*" because in December 2023, to bridge this gap, a paper from Lingmin Ran et al, called *X-Adapter: Adding Universal Compatibility of Plugins for Upgraded Diffusion Model* was published [8]. This paper details an adapter that enables us to use SD V1.5 LoRA and ControlNet in a new SDXL model.

Next, let's start using ControlNet with SD models.

Usage of ControlNet

Before diving into the backend of ControlNet, in this section, we will start using ControlNet to help control image generation.

In the following example, we will first generate an image using SD, take the Canny shape of the object, and then use the Canny shape to generate a new image with the help of ControlNet.

> **Note**
>
> A Canny image refers to an image that has undergone Canny edge detection, which is a popular edge detection algorithm. It was developed by John F. Canny in 1986. [7]

Let's use SD to generate an image using the following code:

1. Generate a sample image using SD:

```
import torch
from diffusers import StableDiffusionPipeline

# load model
text2img_pipe = StableDiffusionPipeline.from_pretrained(
    "stablediffusionapi/deliberate-v2",
     torch_dtype = torch.float16
).to("cuda:0")

# generate sample image
prompt = """
high resolution photo,best quality, masterpiece, 8k
A cute cat stand on the tree branch, depth of field, detailed
body
"""

neg_prompt = """
paintings,ketches, worst quality, low quality, normal quality,
lowres,
monochrome, grayscale
"""

image = text2img_pipe(
    prompt = prompt,
    negative_prompt = neg_prompt,
    generator = torch.Generator("cuda").manual_seed(7)
).images[0]
image
```

We will see an image of a cat, as shown in *Figure 13.1*:

Figure 13.1: A cat, generated by SD

2. Then we will get the Canny shape of the sample image.

    We will need another package, `controlnet_aux`, to create a Canny image from an image. Simply execute the following two lines of `pip` commands to install `controlnet_aux`:

    ```
    pip install opencv-contrib-python
    pip install controlnet_aux
    ```

    We can generate the image Canny edge shape with three lines of code:

    ```
    from controlnet_aux import CannyDetector
    canny = CannyDetector()
    image_canny = canny(image, 30, 100)
    ```

    Here's a breakdown of the code:

    *   `from controlnet_aux import CannyDetector`: This line imports the `CannyDetector` class from the `controlnet_aux` module. There are many other detectors.

    *   `image_canny = canny(image, 30, 100)`: This line calls the `__call__` method of the `CannyDetector` class (which is implemented as a callable object) with the following arguments:

        *   `image`: This is the input image to which the Canny edge detection algorithm will be applied.

        *   `30`: This is the lower threshold value for the edges. Any edges with an intensity gradient below this value will be discarded.

        *   `100`: This is the upper threshold value for the edges. Any edges with an intensity gradient above this value will be considered strong edges.

The preceding code will generate the Canny image shown in *Figure 13.2*:

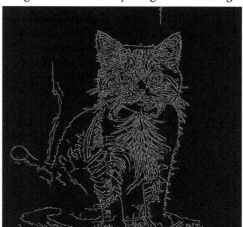

Figure 13.2: Canny image of a cat

3. We will now use the ControlNet model to generate a new image based on this Canny image.
    First, let's load up the ControlNet model:

```
from diffusers import ControlNetModel
canny_controlnet = ControlNetModel.from_pretrained(
    'takuma104/control_v11',
    subfolder='control_v11p_sd15_canny',
    torch_dtype=torch.float16
)
```

The code will download the ControlNet model from Hugging Face automatically at your first run.
If you have the ControlNet `safetensors` model in your storage and want to use your own
model, you can convert the file to the diffuser format first. You can find the conversion code in
Chapter 6. Then, replace `takuma104/control_v11` with the path to the ControlNet model.

4. Initialize a ControlNet pipeline:

```
from diffusers import StableDiffusionControlNetImg2ImgPipeline
cn_pipe = \
    StableDiffusionControlNetImg2ImgPipeline.from_pretrained(
    "stablediffusionapi/deliberate-v2",
    torch_dtype = torch.float16,
    controlnet = canny_controlnet
)
```

Note that you can freely swap `stablediffusionapi/deliberate-v2` with any other
SD v1.5 models from the community.

5. Generate the new image using the ControlNet pipeline. In the following example, we will replace the cat with a dog:

```
prompt = """
high resolution photo,best quality, masterpiece, 8k
A cute dog stand on the tree branch, depth of field, detailed
body
"""

neg_prompt = """
paintings,ketches, worst quality, low quality, normal quality,
lowres,
monochrome, grayscale
"""
image_from_canny = single_cn_pipe(
    prompt = prompt,
    negative_prompt = neg_prompt,
    image = canny_image,
    generator = torch.Generator("cuda").manual_seed(2),
    num_inference_steps = 30,
    guidance_scale = 6.0
).images[0]
image_from_canny
```

These lines of code will generate a new image following the Canny edge, but the little cat is now a dog, as shown in *Figure 13.3*:

Figure 13.3: A dog, generated using the cat Canny image with ControlNet

The cat's body structure and shape are preserved. Feel free to change the prompt and settings to explore the amazing capabilities of the model. One thing to note is that if you don't provide a prompt to the ControlNet pipeline, the pipeline will still output a meaningful image, maybe another style of cat, which means the ControlNet model learned the underlying meaning of a certain Canny edge.

In this example, we used only one ControlNet model, but we can also provide multiple ControlNet models to one pipeline.

Using multiple ControlNets in one pipeline

In this section, we will initialize one more ControlNet, NormalBAE, and then feed the Canny and NormalBAE ControlNet models together to form a pipeline.

Let's generate a Normal BAE as one additional control image. Normal BAE is a model that's used to estimate a normal map using the normal uncertainty method [4] proposed by Bae et al:

```python
from controlnet_aux import NormalBaeDetector
normal_bae = \
    NormalBaeDetector.from_pretrained("lllyasviel/Annotators")
image_canny = normal_bae(image)
image_canny
```

This code will generate the original image's Normal BAE map, as shown in *Figure 13.4*:

Figure 13.4: Normal BAE image of the generated cat

Now, let's initialize two ControlNet models for one pipeline: one Canny ControlNet model, and another NormalBae ControlNet model:

```python
from diffusers import ControlNetModel
canny_controlnet = ControlNetModel.from_pretrained(
    'takuma104/control_v11',
    subfolder='control_v11p_sd15_canny',
    torch_dtype=torch.float16
)
bae_controlnet = ControlNetModel.from_pretrained(
```

```
    'takuma104/control_v11',
    subfolder='control_v11p_sd15_normalbae',
    torch_dtype=torch.float16
)
controlnets = [canny_controlnet, bae_controlnet]
```

From the code, we can easily see that all ControlNet models share the same architecture. To load different ControlNet models, we only need to change the model name. Also, note that the two ControlNet models are in a Python `controlnets` `list`. We can provide these ControlNet models to the pipeline directly, as shown here:

```
from diffusers import StableDiffusionControlNetPipeline
two_cn_pipe = StableDiffusionControlNetPipeline.from_pretrained(
    "stablediffusionapi/deliberate-v2",
    torch_dtype = torch.float16,
    controlnet = controlnets
).to("cuda")
```

In the inference stage, use one additional parameter, `controlnet_conditioning_scale`, to control the influence scale of each ControlNet:

```
prompt = """
high resolution photo,best quality, masterpiece, 8k
A cute dog on the tree branch, depth of field, detailed body,
"""

neg_prompt = """
paintings,ketches, worst quality, low quality, normal quality, lowres,
monochrome, grayscale
"""
image_from_2cn = two_cn_pipe(
    prompt = prompt,
    image = [canny_image,bae_image],
    controlnet_conditioning_scale = [0.5,0.5],
    generator = torch.Generator("cuda").manual_seed(2),
    num_inference_steps = 30,
    guidance_scale = 5.5
).images[0]
image_from_2cn
```

This code will give us another image, as shown in *Figure 13.5*:

Figure 13.5: A dog generated from Canny ControlNet and a Normal BAE ControlNet

In `controlnet_conditioning_scale = [0.5,0.5]`, I give each ControlNet model a `0.5` scale value. The two scale values add up to `1.0`. We should give weights that add up to no more than 2. Values that are too will lead to undesired images. For instance, if you give weights of `1.2` and `1.3` to each ControlNet model, like `controlnet_conditioning_scale = [1.2,1.3]`, you may get an undesired image.

If we have successfully generated images using the ControlNet models, we have together witnessed the power of ControlNet. In the next section, we will discuss how ControlNet works.

How ControlNet works

In this section, we will drill down into the ControlNet structure and see how ControlNet works internally.

ControlNet works by injecting additional conditions into the blocks of a neural network. As shown in *Figure 13.6*, the trainable copy is the ControlNet block that adds additional guidance to the original SD UNet block:

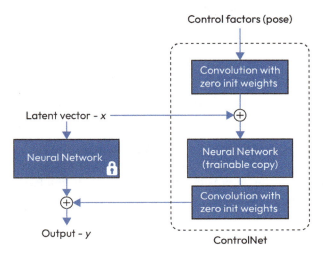

Figure 13.6: Adding ControlNet components

During the training stage, we take a copy of the target layer block as the ControlNet block. In *Figure 13.6*, it is denoted as a **trainable copy**. Unlike typical neural network initialization with Gaussian distributions for all parameters, ControlNet utilizes pre-trained weights from the Stable Diffusion base model. Most of these base model parameters are frozen (with the option to unfreeze them later) and only the additional ControlNet components are trained from scratch.

During training and inference, the input x is usually a 3D dimensional vector, $x \in \mathbb{R}^{h \times w \times c}$, with h, w, c as the height, width, and number of channels. c is a conditioning vector that we will pass into the SD UNet and also to the ControlNet model network.

The **zero convolution** plays a pivotal role in this process. **Zero convolutions** are 1D convolutions with weights and biases initialized to zero. The advantage of zero convolution is that, even without a single training step, the value injected from ControlNet will have no effect on image generation. This ensures that the side network does not negatively impact image generation at any stage.

You might be thinking: if the weight of a convolution layer is zero, wouldn't the gradient also be zero, rendering the network unable to learn? However, as the paper's authors explain [5], the reality is more nuanced.

Let's consider one simple case:

$$y = wx + b$$

Then we have the following:

$$\partial y / \partial w = x, \partial y / \partial x = w, \partial y / \partial b = 1$$

And if $w = 0$ and $x \neq 0$, then we have this:

$$\partial y / \partial w \neq 0, \partial y / \partial x = 0, \partial y / \partial b \neq 0$$

This means as long as $x \neq 0$, one gradient descent iteration will make w non-zero. Then, we have:

$$\partial y / \partial x \neq 0$$

So, the zero convolutions will progressively become a common convolution layer with non-zero weights. What a genius design!

The SD UNet is only connected with a ControlNet at the encoder blocks and the middle blocks. The trainable blue blocks and the white zero convolution layers are added to build a ControlNet. It's simple and effective.

In the original paper— Adding Conditional Control to Text-to-Image Diffusion Models [1] by Lvmin Zhang et al — its authors also provided an ablative study and discussed lots of different cases, such as swapping the **zero convolution** layer with a **traditional convolution** layer and comparing the differences. It is a great paper and fun to read.

Further usage

In this section, we will introduce more usage of ControlNet, covering SD V1.5 and also SDXL.

More ControlNets with SD

The author of the ControlNet-v1-1-nightly repository [3] lists all the currently available V1.1 ControlNet models for SD. As of the time I am writing this chapter, the list is as follows:

```
control_v11p_sd15_canny
control_v11p_sd15_mlsd
control_v11f1p_sd15_depth
control_v11p_sd15_normalbae
control_v11p_sd15_seg
control_v11p_sd15_inpaint
control_v11p_sd15_lineart
control_v11p_sd15s2_lineart_anime
control_v11p_sd15_openpose
control_v11p_sd15_scribble
control_v11p_sd15_softedge
control_v11e_sd15_shuffle
control_v11e_sd15_ip2p
control_v11f1e_sd15_tile
```

You can simply swap the ControlNet model's name with one from this list to start using it. Generate the control image using one of the annotators from the open source ControlNet auxiliary models[6].

Considering the speed of development in the field of AI, when you are reading this, the version may have increased to v1.1+. However, the underlying mechanism should be the same.

SDXL ControlNets

As I am writing this chapter, SDXL has just been released, and this new model generates excellent images with shorter prompts than before. The Hugging Face Diffusers team trained and provided several ControlNet models for the XL models. Its usage is almost the same as the previous version. Here, let's use the `controlnet-openpose-sdxl-1.0` open pose ControlNet for SDXL.

Note that you will need a dedicated GPU with more than 15 GB of VRAM to run the following example.

Let's initialize an SDXL pipeline using the following code:

```
import torch
from diffusers import StableDiffusionXLPipeline
sdxl_pipe = StableDiffusionXLPipeline.from_pretrained(
    "RunDiffusion/RunDiffusion-XL-Beta",
```

```
    torch_dtype = torch.float16,
    load_safety_checker = False
)
sdxl_pipe.watermark = None
```

Then, generate an image with a man in it:

```
from diffusers import EulerDiscreteScheduler
prompt = """
full body photo of young man, arms spread
white blank background,
glamour photography,
upper body wears shirt,
wears suit pants,
wears leather shoes
"""
neg_prompt = """
worst quality,low quality, paint, cg, spots, bad hands,
three hands, noise, blur, bad anatomy, low resolution, blur face, bad
face
"""
sdxl_pipe.to("cuda")

sdxl_pipe.scheduler = EulerDiscreteScheduler.from_config(
    sdxl_pipe.scheduler.config)
image = sdxl_pipe(
    prompt = prompt,
    negative_prompt = neg_prompt,
    width = 832,
    height = 1216
).images[0]
sdxl_pipe.to("cpu")
torch.cuda.empty_cache()
image
```

The code will generate an image, as shown in *Figure 13.7*:

Figure 13.7: A man in a suit, generated by SDXL

We can use `OpenposeDetector` from `controlnet_aux` [6] to extract the pose:

```
from controlnet_aux import OpenposeDetector
open_pose = \
    OpenposeDetector.from_pretrained("lllyasviel/Annotators")
pose = open_pose(image)
pose
```

We will get the pose image shown in *Figure 13.8*:

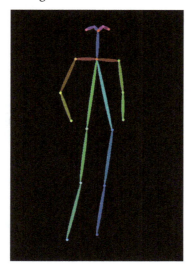

Figure 13.8: Pose image of the man in a suit

Now, let's start an SDXL pipeline with the SDXL ControlNet open pose model:

```
from diffusers import StableDiffusionXLControlNetPipeline
from diffusers import ControlNetModel
sdxl_pose_controlnet = ControlNetModel.from_pretrained(
    "thibaud/controlnet-openpose-sdxl-1.0",
    torch_dtype=torch.float16,
)

sdxl_cn_pipe = StableDiffusionXLControlNetPipeline.from_pretrained(
    "RunDiffusion/RunDiffusion-XL-Beta",
    torch_dtype = torch.float16,
    load_safety_checker = False,
    add_watermarker = False,
    controlnet = sdxl_pose_controlnet
)
sdxl_cn_pipe.watermark = None
```

Now we can use the new ControlNet pipeline to generate a new image from the pose image with the same style. We will reuse the prompt but replace **man** with **woman**. We are aiming to generate a new image of a woman in a suit but in the same pose as the previous image of a man:

```
from diffusers import EulerDiscreteScheduler
prompt = """
full body photo of young woman, arms spread
white blank background,
glamour photography,
wear sunglass,
upper body wears shirt,
wears suit pants,
wears leather shoes
"""
neg_prompt = """
worst quality,low quality, paint, cg, spots, bad hands,
three hands, noise, blur, bad anatomy, low resolution,
blur face, bad face
"""
sdxl_cn_pipe.to("cuda")

sdxl_cn_pipe.scheduler = EulerDiscreteScheduler.from_config(
    sdxl_cn_pipe.scheduler.config)
generator = torch.Generator("cuda").manual_seed(2)

image = sdxl_cn_pipe(
    prompt = prompt,
```

```
    negative_prompt = neg_prompt,
    width = 832,
    height = 1216,
    image = pose,
    generator = generator,
    controlnet_conditioning_scale = 0.5,
    num_inference_steps = 30,
    guidance_scale = 6.0
).images[0]
sdxl_cn_pipe.to("cpu")
torch.cuda.empty_cache()
image
```

The code generates a new image with the same pose, exactly matching the expectations, as shown in *Figure 13.9*:

Figure 13.9: A woman in a suit, generated using an SDXL ControlNet

We will further discuss Stable Diffusion XL in *Chapter 16*.

Summary

In this chapter, we introduced a way to precisely control image generation using SD ControlNets. From the detailed samples we have provided, you can start using one or multiple ControlNet models with SD v1.5 and also SDXL.

We also drilled down into the internals of ControlNet, explaining how it works in a nutshell.

We can use ControlNet in lots of applications, including applying a style to an image, applying a shape to an image, merging two images into one, and generating a human body using a posed image. It is powerful and amazingly useful in many ways. Our imagination is the only limitation.

However, there is one other limitation: it is hard to align the background and overall context between two generations (with different seeds). You may want to use ControlNet to generate a video from the extracted frames from a source video, but the results are still not ideal.

In the next chapter, we will cover a solution to generate video and animation using SD.

References

1. Adding conditional control to text-to-image diffusion models: `https://arxiv.org/abs/2302.05543`

2. ControlNet v1.0 GitHub repository: `https://github.com/lllyasviel/ControlNet`

3. ControlNet v1.1 GitHub repository: `https://github.com/lllyasviel/ControlNet-v1-1-nightly`

4. `surface_normal_uncertainty`: `https://github.com/baegwangbin/surface_normal_uncertainty`

5. Zero convolution FAQ: `https://github.com/lllyasviel/ControlNet/blob/main/docs/faq.md`

6. ControlNet AUX: `https://github.com/patrickvonplaten/controlnet_aux`

7. Canny edge detector: `https://en.wikipedia.org/wiki/Canny_edge_detector`

8. X-Adapter: Adding Universal Compatibility of Plugins for Upgraded Diffusion Model: `https://showlab.github.io/X-Adapter/`

14

Generating Video Using Stable Diffusion

Harnessing the power of the Stable Diffusion model, we can generate high-quality images using techniques such as LoRA, text embedding, and ControlNet. A natural progression from static images is toward dynamic content, that is, videos. Can we generate consistent videos using the Stable Diffusion model?

The Stable Diffusion model's UNet architecture, while effective for single-image processing, lacks contextual awareness when dealing with multiple images. Consequently, generating identical or consistently related images with the same prompt and parameters but different seeds is challenging. The resulting images may vary significantly in color, shape, or style due to the randomness introduced by the model's nature.

One might consider an image-to-image pipeline or a ControlNet approach, where a video clip is segmented into individual images, and each image is processed sequentially. However, maintaining consistency across the entire sequence, especially when applying significant changes (such as transforming a realistic video into a cartoon), remains a challenge. Even with pose alignment, the output video may still exhibit noticeable flickering.

A breakthrough came with the publication of *AnimateDiff: Animating Your Personalized Text-to-Image Diffusion Models without Specific Tuning* [1] by Yuwei Gao and colleagues. This work paved the way for generating consistent images from text, thereby enabling the creation of short videos.

In this chapter, we will explore the following:

- The principles of text-to-video generation
- Practical applications of AnimateDiff
- Utilizing Motion LoRA to control animation motion

By the end of this chapter, you will understand the theoretical aspects of video generation, the inner workings of AnimateDiff, and why this methodology is effective in creating consistent and coherent images. With the provided sample code, you will be able to generate a 16-frame video. You can then apply Motion LoRA to manipulate the animation's motion.

Please note that the results of this chapter cannot be fully appreciated in a static format like paper or PDF. For the best experience, we encourage you to engage with the associated sample code, run it, and observe the generated video.

Technical requirements

In this chapter, we will employ `AnimateDiffPipeline`, available in the `Diffusers` library, to generate videos. You won't need to install any extra tools or packages, as Diffusers (after version 0.23.0) offers all the required components and classes. Throughout the chapter, I will guide you through the usage of these features.

To export the result in MP4 video format, you will also need to install the `opencv-python` package:

```
pip install opencv-python
```

Also, note that the `AnimateDiffPipeline` will require at least 8 GB of VRAM to generate a 16-frame 256x256 video clip.

The principles of text-to-video generation

The Stable Diffusion UNet, while effective for generating single images, falls short when it comes to generating consistent images due to its lack of contextual awareness. Researchers have proposed solutions to overcome this limitation, such as incorporating temporal information from the preceding one or two frames. However, this approach still fails to ensure pixel-level consistency, leading to noticeable differences between consecutive images and flickering in the generated video.

To address this inconsistency problem, the authors of AnimateDiff trained a separated motion model – a zero-initialized convolution side model – similar to the ControlNet model. Further, rather than controlling an image, the motion model is applied to a series of continuous frames, as shown in *Figure 14.1*:

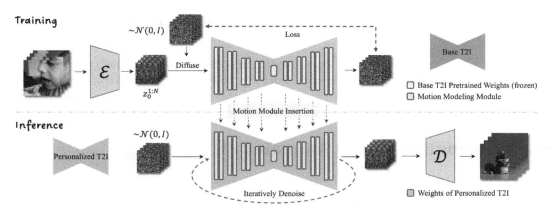

Figure 14.1: The architecture of AnimatedDiff

The process involves training a motion modeling module on video datasets to extract motion priors while keeping the base Stable Diffusion model frozen. Motion priors are the prior knowledge about motion in order to guide the generation or customization of videos. During the training stage, a **motion module** (also called **Motion UNet**) is added to the Stable Diffusion UNet. Similar to normal Stable Diffusion V1.5 UNet training, this Motion UNet will work on all frames simultaneously. We can treat them as images in one batch from the same video clip.

For instance, if we feed in a video with 16 frames, the motion module with attention headers will be trained to consider all 16 frames. If we look into the implementation source code, `TransformerTemporalModel` [4] is the core component of `MotionModules` [3].

During the inference, the time when we want to generate videos, the motion module will be loaded and its weights will be merged into Stable Diffusion UNet. When we want to generate a video with 16 frames, the pipeline will first initialize 16 random latents with Gaussian noise – $\mathcal{N}(0,1)$. Without the motion module, the Stable Diffusion UNet will remove noise and generate 16 independent images. However, with the help of the motion module with the Transformer attention header built inside, the motion UNet attempts to create 16 correlated frames. You may ask, why are the images correlated? That is because the frames in the training video are correlated. After the denoising stage, the decoder $\mathcal{D}$ from the VAE will convert the 16 latents into pixel images.

The Motion UNet is responsible for introducing correlations between successive frames in the generated video. It is similar to the correlation of different areas in one image. This is because the attention header pays attention to different parts of the image, and the model learned this knowledge during the training stage. Likewise, during the video generation, the model learned the correlation between frames during the training stage.

At its core, this approach involves designing an attention mechanism that operates on a sequence of continuous images. By learning the relationships between frames, AnimateDiff can generate more consistent and coherent images from text. Furthermore, since the base Stable Diffusion model remains locked, various Stable Diffusion extension techniques, such as LoRA, textual embedding, ControlNet, and image-to-image generation, can be applied to AnimateDiff as well.

Anything that works for standard Stable Diffusion, in theory, should also work for AnimateDiff to AnimateDiff as well!

Before moving on to the next section, be aware that the AnimateDiff model requires a minimum of 12 GB of VRAM to generate a 16-frame, 256x256 video clip. To truly grasp the concept, writing code to utilize AnimateDiff is highly recommended. Now, let's proceed to generate a short video (in GIF and MP4 format) using AnimateDiff.

Practical applications of AnimateDiff

The original AnimateDiff code and model were released as a standalone GitHub repository [2]. While the author provided sample code and Google Colab to demonstrate the results, users still needed to manually pull the code and download the model file to use it, being cautious about package versions.

In November 2023, Dhruv Nair [9] merged an AnimateDiff Pipeline for Diffusers, allowing users to generate video clips using the AnimateDiff pertained model without leaving the `Diffusers` package. Here's how to use the AnimatedDiff pipeline from Diffusers:

1. Install this specific version of Diffusers with the integrated AnimateDiff code:

    ```
    pip install diffusers==0.23.0
    ```

    At the time of writing this chapter, the version of Diffusers with the latest AnimateDiff code is 0.23.0. By specifying this version number, you can ensure that the sample code runs smoothly and error-free, as it was tested against this particular version.

    You can also try installing the latest version of Diffusers, as it may have added more features to the pipeline by the time you read this:

    ```
    pip install -U diffusers
    ```

2. Load up the motion adapter. We will use the pre-trained motion adapter model from the author of the original paper:

    ```
    from diffusers import MotionAdapter
    adapter = MotionAdapter.from_pretrained(
        "guoyww/animatediff-motion-adapter-v1-5-2"
    )
    ```

3. Load up an AnimateDiff pipeline from a Stable Diffusion v1.5-based checkpoint model:

    ```
    from diffusers import AnimateDiffPipeline
    pipe = AnimateDiffPipeline.from_pretrained(
        "digiplay/majicMIX_realistic_v6",
        motion_adapter    = adapterm,
        safety_checker    = None
    )
    ```

4. Use a proper scheduler. The scheduler plays an important role in the process of generating coherent images. An comparative study conducted by the author of the paper shows different schedulers can lead to different results. Experimentation shows that the `EulerAncestralDiscreteScheduler` scheduler with the following setting can generate relatively good results:

```
from diffusers import EulerAncestralDiscreteScheduler
scheduler = EulerAncestralDiscreteScheduler.from_pretrained(
    model_path,
    subfolder        = "scheduler",
    clip_sample      = False,
    timestep_spacing = "linspace",
    steps_offset     = 1
)
pipe.scheduler = scheduler
pipe.enable_vae_slicing()
pipe.enable_model_cpu_offload()
```

To optimize VRAM usage, you can employ two strategies. First, use `pipe.enable_vae_slicing()` to configure the VAE to decode one frame at a time, thereby reducing memory consumption. Additionally, utilize `pipe.enable_model_cpu_offload()` to offload idle sub-models to the CPU, further decreasing VRAM usage.

5. Generate coherent images:

```
import torch
from diffusers.utils import export_to_gif, export_to_video

prompt = """photorealistic, 1girl, dramatic lighting"""

neg_prompt = """worst quality, low quality, normal quality,
lowres, bad anatomy, bad hands, monochrome, grayscale watermark,
moles"""
#pipe.to("cuda:0")

output = pipe(
    prompt = prompt,
    negative_prompt = neg_prompt,
    height = 256,
    width = 256,
    num_frames = 16,
    num_inference_steps = 30,
    guidance_scale= 8.5,
    generator = torch.Generator("cuda").manual_seed(7)
)
```

```
frames = output.frames[0]
torch.cuda.empty_cache()

export_to_gif(frames, "animation_origin_256_wo_lora.gif")
export_to_video(frames, "animation_origin_256_wo_lora.mp4")
```

Now, you should be able to see a GIF file generated using the 16 frames produced by AnimateDiff. This GIF uses 16 256x256 images. You can apply the image super-resolution techniques introduced in *Chapter 11* to upscale the image and create a 512x512 GIF. I will not duplicate the code in this chapter. It is highly recommended to leverage the skills learned in *Chapter 11* to further enhance the quality of video generation.

Utilizing Motion LoRA to control animation motion

Besides the motion adapter model, the author of the paper also introduced Motion LoRA to control the motion style. Motion LoRA is the same LoRA adapter we introduced in *Chapter 8*. As mentioned before, the AnimateDiff pipeline supports all other community-shared LoRAs. You can find these Motion LoRAs on the author's Hugging Face repository [8].

These Motion LoRAs can be used to control the camera view. Here, we will use zoom-in LoRA – guoyww/animatediff-motion-lora-zoom-in – as an example. The zoom-in will guide the model to generate a video with zoom-in motion.

The usage is simply one additional line of code:

```
pipe.load_lora_weights("guoyww/animatediff-motion-lora-zoom-in",
    adapter_name="zoom-in")
```

Here is the complete generation code. We are mostly reusing the code from the previous section:

```
import torch
from diffusers.utils import export_to_gif, export_to_video

prompt = """
photorealistic, 1girl, dramatic lighting
"""
neg_prompt = """
worst quality, low quality, normal quality, lowres, bad anatomy, bad
hands
, monochrome, grayscale watermark, moles
"""
pipe.to("cuda:0")

pipe.load_lora_weights("guoyww/animatediff-motion-lora-zoom-in",
adapter_name="zoom-in")
```

```
output = pipe(
    prompt = prompt,
    negative_prompt = neg_prompt,
    height = 256,
    width = 256,
    num_frames = 16,
    num_inference_steps = 40,
    guidance_scale = 8.5,
    generator = torch.Generator("cuda").manual_seed(123)
)
frames = output.frames[0]

pipe.to("cpu")
torch.cuda.empty_cache()

export_to_gif(frames, "animation_origin_256_w_lora_zoom_in.gif")
export_to_video(frames, "animation_origin_256_w_lora_zoom_in.mp4")
```

You should see a zoom-in GIF clip is generated under the same folder, named `animation_origin_256_w_lora_zoom_in.gif` and an MP4 video clip is generated named `animation_origin_256_w_lora_zoom_in.mp4`.

Summary

Every day, the quality and duration of text-to-video samples circulating on social networks are improving. It's likely that by the time you read this chapter, the function of the technologies metioned in this chapter will have surpassed what was described here. However, one constant is the concept of training a model to apply an attention mechanism to a sequence of images.

At the time of writing, OpenAI's Sora [9] has just been released. This technology can generate high-quality videos based on the Transformer Diffusion architecture. This is a similar methodology to that used in AnimatedDiff, which combines the Transformer and diffusion models.

What sets AnimatedDiff apart is its openness and adaptability. It can be applied to any community model with the same base checkpoint version, a feature not currently offered by any other solution. Furthermore, the authors of the paper have completely open-sourced the code and model.

This chapter primarily discussed the challenges of text-to-image generation, then introduced AnimatedDiff, explaining how and why it works. We also provided a sample code to use the AnimatedDiff pipeline from the Diffusers package to generate a GIF clip from 16 coherent images on your own GPU.

In the next chapter, we will explore the solutions for generating text descriptions from an image.

References

1. Yuwei Guo, Ceyuan Yang, Anyi Rao, Yaohui Wang, Yu Qiao, Dahua Lin, and Bo Dai, *AnimateDiff: Animate Your Personalized Text-to-Image Diffusion Models without Specific Tuning*: https://arxiv.org/abs/2307.04725

2. Original AnimateDiff code repository: https://github.com/guoyww/AnimateDiff

3. Diffusers Motion modules implementation: https://github.com/huggingface/diffusers/blob/3dd4168d4c96c429d2b74c2baaee0678c57578da/src/diffusers/models/unets/unet_motion_model.py#L50

4. Hugging Face Diffusers TransformerTemporalModel implementation: https://github.com/huggingface/diffusers/blob/3dd4168d4c96c429d2b74c2baae-e0678c57578da/src/diffusers/models/transformers/transformer_temporal.py#L41

5. [4] Dhruv Nair, https://github.com/DN6

6. AnimateDiff proposal pull request: https://github.com/huggingface/diffusers/pull/5413

7. animatediff-motion-adapter-v1-5-2: https://huggingface.co/guoyww/animatediff-motion-adapter-v1-5-2

8. Yuwei Guo's Hugging Face repository: https://huggingface.co/guoyww

9. Video generation models as world simulators: https://openai.com/research/video-generation-models-as-world-simulators

15

Generating Image Descriptions Using BLIP-2 and LLaVA

Imagine you have an image in hand and need to upscale it or generate new images based on it, but you don't have the prompt or description associated with it. You may say, *"Fine, I can write up a new prompt for it."* For one image, that is acceptable, what if there are thousands or even millions of images without descriptions? It is impossible to write them all up manually.

Fortunately, we can use artificial intelligence (AI) to help us generate descriptions. There are many pretrained models that can achieve this goal, and the number is always increasing. In this chapter, I am going to introduce two AI solutions to generate the caption, description, or prompt for an image, all fully automated:

- BLIP-2: Bootstrapping Language-Image Pre-training with Frozen Image Encoders and Large Language Models [1]
- LLaVA: Large Language and Vision Assistant [3]

BLIP-2 [1] is fast and requires relatively low hardware, while LLaVA [3] (with its `llava-v1.5-13b` model) is the newest and most powerful model at the time of writing.

By the end of this chapter, you will be able to do the following:

- Generally understand how BLIP-2 and LLaVA work
- Write up Python code to use BLIP-2 and LLaVA to generate descriptions from images

Technical requirements

Before diving into the BLIP-2 and LLaVA, let's use Stable Diffusion to generate an image for testing.

First, load up a `deliberate-v2` model without sending it to CUDA:

```
import torch
from diffusers import StableDiffusionPipeline
text2img_pipe = StableDiffusionPipeline.from_pretrained(
    "stablediffusionapi/deliberate-v2",
    torch_dtype = torch.float16
)
```

Next, in the following code, we first send the model to CUDA and generate an image, then we offload the model to CPU RAM, and clear the model out from CUDA:

```
text2img_pipe.to("cuda:0")
prompt ="high resolution, a photograph of an astronaut riding a horse"
input_image = text2img_pipe(
    prompt = prompt,
    generator = torch.Generator("cuda:0").manual_seed(100),
    height = 512,
    width = 768
).images[0]
text2img_pipe.to("cpu")
torch.cuda.empty_cache()
input_image
```

The preceding code will give us an image similar to that shown in the following figure, which will be used in the following sections:

Figure 15.1: An image of an astronaut riding a horse, generated by SD v1.5

Now, let's get started.

BLIP-2 – Bootstrapping Language-Image Pre-training

In the *BLIP: Bootstrapping Language-Image Pre-training for Unified Vision-Language Understanding and Generation* paper [4], Junnan Li et al. proposed a solution to bridge the gap between natural language and vision modalities. Notably, the BLIP model has demonstrated exceptional capabilities in generating high-quality image descriptions, surpassing existing benchmarks at the time of its publication.

The reason behind its excellent quality is that Junnan Li et al. used an innovative technique to build two models from their first pretrained model:

- Filter model
- Captioner model

The filter model can filter out low-quality text-image pairs, thus improving the training data quality, while its caption generation model can generate surprisingly good, short descriptions for the image. With the help of these two models, the authors of the paper not only improved the training data quality but also enlarged its size automatically. Then, they used the boosted train data to train the BLIP model again and the result was impressively good. But this was the story of 2022.

In June 2023, the same team from Salesforce brought out the new BLIP-2.

How BLIP-2 works

BLIP was good at the time, but the language part of its model was still relatively weak. **Large language models (LLMs)** such as OpenAI's GPT and Meta's LLaMA are powerful, but also extremely expensive to train. So the BLIP team framed the challenge by asking themselves: can we take off-the-shelf, pretrained frozen image encoders and frozen LLMs and use them for vision language pretraining while still preserving their learned representations?

The answer is yes. BLIP-2 solved this by introducing a Query Transformer that helps generate the visual representations corresponding to a text caption, which then is fed to a frozen LLM to decode text descriptions.

The Query Transformer, often referred to as the Q-Former [2], is a crucial component of the BLIP-2 model. It serves as a bridge connecting the frozen image encoder and the frozen LLM. The primary function of the Q-Former is to map a set of "query tokens" to query embeddings. These query embeddings help in extracting visual features from the image encoder that are most pertinent to the given text instruction.

During the training process of the BLIP-2 model, the weights of the image encoder and the LLM remain frozen. Meanwhile, the Q-Former undergoes training, allowing it to adapt and optimize its performance based on the specific task requirements. By employing a set of learnable query vectors, the Q-Former effectively distills valuable information from the image encoder, making it possible for the LLM to generate accurate and contextually appropriate responses grounded in visual content.

A similar concept is also employed in LLaVA, which we will discuss later. The core idea of BLIP is to reuse the effective vision and language components, and only train a middle model to bridge them together.

Next, let's start using BLIP-2.

Using BLIP-2 to generate descriptions

Using BLIP-2 is easy and clean with the help of the Hugging Face transformers [5] package. If you don't have the package installed, simply run the following command to install or update it to the newest version:

```
pip install -U transformer
```

Then load up the BLIP-2 model data with the following code:

```
from transformers import AutoProcessor, Blip2ForConditionalGeneration
import torch

processor = AutoProcessor.from_pretrained("Salesforce/blip2-opt-2.7b")
# by default `from_pretrained` loads the weights in float32
# we load in float16 instead to save memory
device = "cuda" if torch.cuda.is_available() else "cpu"
model = Blip2ForConditionalGeneration.from_pretrained(
    "Salesforce/blip2-opt-2.7b",
    torch_dtype=torch.float16
).to(device)
```

Your first run will automatically download the model weights data from the Hugging Face model repository. It may take some time, so please be patient. Once the downloading is finished, run the following code to ask BLIP-2 about the image we provide:

```
prompt = "describe the content of the image:"
inputs = processor(
    input_image,
    text=prompt,
    return_tensors="pt"
).to(device, torch.float16)

generated_ids = model.generate(**inputs, max_new_tokens=768)
generated_text = processor.batch_decode(
    generated_ids,
    skip_special_tokens=True
)[0].strip()
print(generated_text)
```

The code returns the description `astronaut on horseback in space`, which is good and accurate. What if we ask how many planets are in the background? Let's change the prompt to `how many planets in the background:`. And it returns `the universe is bigger than you think`. Not good enough this time.

So, BLIP-2 is good at generating short descriptions of an entire image quickly. However, to generate more detailed descriptions or even interact with images, we can leverage the power of LLaVA.

LLaVA – Large Language and Vision Assistant

As suggested by its name **LLaVA** [3], this model is very close to LLaMA, not only in name but also in terms of their internals. LLaVA uses LLaMA as its language part. making it possible to swap out the language model if needed This is definitely a killer feature for many scenarios. One of the key features of Stable Diffusion is its openness for model swapping and fine-tuning. Similar to Stable Diffusion, LLaVA is designed to leverage open-sourced LLM models.

Next, let's take a look at how LLaVA works.

How LLaVA works

The LLaVA authors, Haotian Liu et al. [3], present a beautiful, accurate diagram showing how the model leverages pretrained CLIP and LLaMA models in its architecture, as shown in the following figure:

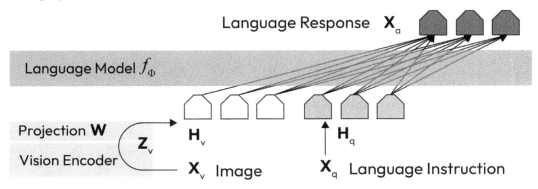

Figure 15.2: Architecture of LLaVA

Let's read the diagram from the bottom up. During the inference, we provide an image denoted as X_v, and a language instruction denoted as X_q. The vision encoder is the CLIP vision encoder ViT-L/14 [6]. The same CLIP is used by Stable Diffusion v1.5 as its text encoder.

The CLIP model encodes the image to Z_v, and the projection W is the model data provided by LLaVA. The projection model projects the encoded image embedding Z_v to H_v as follows:

$$H_v = W \cdot Z_v$$

On the other side, the language instruction is encoded into CLIP's 512-dimensional embedding chunks. Both the image and language embeddings share the same dimensionality.

In this manner, the language model f_ϕ is aware of both the image and the language! This method bears some similarity to Stable Diffusion's textual inversion technique, being lightweight yet powerful.

Next, let's write some code to instruct LLaVA to interact with an image.

Installing LLaVA

For an optimal experience, it is strongly advised to use LLaVA on a Linux machine. Using it on Windows may result in unexpected missing components. It is also suggested to establish a Python virtual environment for using LLaVA. Detailed steps and commands for setting up a Python virtual environment were provided in *Chapter 2*.

Clone the LLaVA repository to your local folder:

```
git clone https://github.com/haotian-liu/LLaVA.git
cd LLaVA
```

Then install LLaVA by simply running the following command:

```
pip install -U .
```

Next, download the model file from the Hugging Face model repository:

```
# Make sure you have git-lfs installed (https://git-lfs.com)
git lfs install
git clone https://huggingface.co/liuhaotian/llava-v1.5-7b
```

Please note that the model file is large and the download will take some time. At the time of writing this chapter, you can also download the 13B model by simply changing the 7 to 13 in the URL in the preceding code snippet to use the 13B LLaVA model.

That's it for the setup. Now, let's proceed to writing the Python code.

Using LLaVA to generate image descriptions

Since we just installed LLaVA, we can reference the following related modules:

```
from llava.constants import (
    IMAGE_TOKEN_INDEX,
    DEFAULT_IMAGE_TOKEN,
    DEFAULT_IM_START_TOKEN,
    DEFAULT_IM_END_TOKEN
)
```

```
from llava.conversation import (
    conv_templates, SeparatorStyle
)
from llava.model.builder import load_pretrained_model
from llava.mm_utils import (
    process_images,
    tokenizer_image_token,
    get_model_name_from_path,
    KeywordsStoppingCriteria
)
```

Load the `tokenizer`, `image_processor`, and `model` components. `tokenizer` will convert text to token IDs, `image_processor` will convert images to tensors, and `model` is the pipeline that we will use to generate the output:

```
# load up tokenizer, model, image_processor
model_path = "/path/to/llava-v1.5-7b"
model_name = get_model_name_from_path(model_path)
conv_mode = "llava_v1"
tokenizer, model, image_processor, _ = load_pretrained_model(
    model_path = model_path,
    model_base = None,
    model_name = model_name,
    load_4bit = True,
    device = "cuda",
    device_map = {'':torch.cuda.current_device()}
)
```

The following is a breakdown of the preceding code:

- `model_path`: This path points to the folder storing the pretrained models.

- `model_base`: This is set to None, meaning no specific parent architecture has been specified.

- `model_name`: The name of the pretrained model (`llava-v1.5`).

- `load_4bit`: If set to True, this enables 4-bit quantization during inferencing. This reduces memory usage and boosts speed, but might negatively impact the quality of results slightly.

- `device`: This specifies CUDA as the device where computations should occur.

- `device_map`: This is used to map GPU devices to different parts of the model if you want to distribute the workload across multiple GPUs. Since only one device is mapped here, it implies a single GPU execution.

Now, let's create the image descriptions:

1. Create a `conv` object to hold the conversation history:

    ```
    # start a new conversation
    user_input = """Analyze the image in a comprehensive and
    detailed manner"""
    conv = conv_templates[conv_mode].copy()
    ```

2. Convert the image to a tensor:

    ```
    # process image to tensor
    image_tensor = process_images(
        [input_image],
        image_processor,
        {"image_aspect_ratio":"pad"}
    ).to(model.device, dtype=torch.float16)
    ```

3. Append an image placeholder to the conversation:

    ```
    if model.config.mm_use_im_start_end:
        inp = DEFAULT_IM_START_TOKEN + DEFAULT_IMAGE_TOKEN + \
            DEFAULT_IM_END_TOKEN + '\n' + user_input
    else:
        inp = DEFAULT_IMAGE_TOKEN + '\n' + user_input
    conv.append_message(conv.roles[0], inp)
    ```

4. Get the prompt and convert it to tokens for inference:

    ```
    # get the prompt for inference
    conv.append_message(conv.roles[1], None)
    prompt = conv.get_prompt()

    # convert prompt to token ids
    input_ids = tokenizer_image_token(
        prompt,
        tokenizer,
        IMAGE_TOKEN_INDEX,
        return_tensors='pt'
    ).unsqueeze(0).cuda()
    ```

5. Prepare the stopping criteria:

    ```
    stop_str = conv.sep if conv.sep_style != \
        SeparatorStyle.TWO else conv.sep2
    keywords = [stop_str]
    stopping_criteria = KeywordsStoppingCriteria(keywords,
        tokenizer, input_ids)
    ```

6. Finally, get the output from LLaVA:

```
# output the data
with torch.inference_mode():
    output_ids = model.generate(
        input_ids,
        images =image_tensor,
        do_sample = True,
        temperature = 0.2,
        max_new_tokens = 1024,
        streamer = None,
        use_cache = True,
        stopping_criteria = [stopping_criteria]
    )
outputs = tokenizer.decode(output_ids[0,
    input_ids.shape[1]:]).strip()
# make sure the conv object holds all the output
conv.messages[-1][-1] = outputs
print(outputs)
```

As we can see in the following output, LLaVA can generate amazing descriptions:

```
The image features a man dressed in a white space suit, riding
a horse in a desert-like environment. The man appears to be a
space traveler, possibly on a mission or exploring the area. The
horse is galloping, and the man is skillfully riding it.

In the background, there are two moons visible, adding to the
sense of a space-themed setting. The combination of the man in a
space suit, the horse, and the moons creates a captivating and
imaginative scene.</s>
```

I have attempted to condense the code as much as possible, but it remains lengthy. It requires careful copying or transcription into your code editor. I recommend copy-pasting the code provided in the repository that accompanies this book. You can execute the aforementioned code in a single cell to observe how effectively LLaVA generates descriptions from an image.

Summary

In this chapter, our primary focus was on two AI solutions designed to generate image descriptions. The first is BLIP-2, an effective and efficient solution for generating concise captions for images. The second is the LLaVA solution, which is capable of generating more detailed and accurate descriptive information from an image.

With the assistance of LLaVA, we can even interact with an image to extract further information from it.

The integration of vision and language capabilities also lays the groundwork for the development of even more powerful multimodal models, the potential of which we can only begin to imagine.

In the next chapter, let's get started using Stable Diffusion XL.

References

1. Junnan Li, Dongxu Li, Silvio Savarese, Steven Hoi, *BLIP-2: Bootstrapping Language-Image Pre-training with Frozen Image Encoders and Large Language Models*: `https://arxiv.org/abs/2301.12597`

2. BLIP-2 Hugging Face documentation: `https://huggingface.co/docs/transformers/main/model_doc/blip-2`

3. Haotian Liu, Chunyuan Li, Qingyang Wu, Yong Jae Lee, *LLaVA: Large Language and Vision Assistant*: `https://llava-vl.github.io/`

4. Junnan Li, Dongxu Li, Caiming Xiong, Steven Hoi, *BLIP: Bootstrapping Language-Image Pre-training for Unified Vision-Language Understanding and Generation*: `https://arxiv.org/abs/2201.12086`

5. Hugging Face Transformers GitHub repository: `https://github.com/huggingface/transformers`

6. Alec Radford, Jong Wook Kim, Chris Hallacy, Aditya Ramesh, Gabriel Goh, Sandhini Agarwal, Girish Sastry, Amanda Askell, Pamela Mishkin, Jack Clark, Gretchen Krueger, Ilya Sutskever, *Learning transferable visual models from natural language supervision*: `https://arxiv.org/abs/2103.00020`

7. LLaVA GitHub repository: `https://github.com/haotian-liu/LLaVA`

16

Exploring Stable Diffusion XL

After the not-very-successful Stable Diffusion 2.0 and Stable Diffusion 2.1, July 2023 saw the launch of Stability AI's latest release, **Stable Diffusion XL** (**SDXL**) [1]. I eagerly applied the model weights data as soon as registration was open. Both my tests and those conducted by the community indicate that SDXL has made significant strides forward. It now allows us to generate higher-quality images at increased resolutions, vastly outperforming the Stable Diffusion V1.5 base model. Another notable enhancement is the ability to use more intuitive "natural language" prompts to generate images, eliminating the need to cobble together a multitude of "words" to form a meaningful prompt. Furthermore, we can now generate desired images with more concise prompts.

SDXL has improved in almost every aspect compared to the previous versions, and it is worth the time and effort to start using it for better and stable image generation. In this chapter, we will discuss in detail what's new in SDXL and explain why the aforementioned changes led to its improvements. For example, we will explore what is new in the **Variational Autoencoder** (**VAE**), UNet, and TextEncoder design compared to Stable Diffusion V1.5. In a nutshell, this chapter will cover the following:

- What's new in SDXL?
- Using SDXL

Then, we will use Python code to demonstrate the latest SDXL base and community models in action. We will cover basic usage and also advanced usage, such as loading multiple LoRA models and using unlimited weighted prompts.

Let's begin.

What's new in SDXL?

SDXL is still a latent diffusion model, maintaining the same overall architecture used in Stable Diffusion v1.5. According to the original paper behind SDXL [2], SDXL expands every component, making them wider and bigger. The SDXL backbone UNet is three times larger, there are two text encoders in the SDXL base model, and a separate diffusion-based refinement model is included. The overall architecture is shown in *Figure 16.1*:

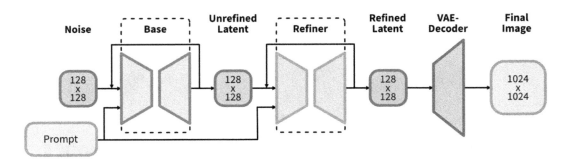

Figure 16.1: SDXL architecture

Note that the refiner is optional; we can decide whether to use the refiner model or not. Next, let's drill down to each component one by one.

The VAE of the SDXL

A **VAE** is a pair of encoder and decoder neural networks. A VAE encoder encodes an image into a latent space, and its paired decoder can decode a latent image to a pixel image. Many articles on the web tell us that a VAE is a technique used to improve the quality of images; however, this is not the whole picture. The core responsibility of VAE in Stable Diffusion is converting pixel images to and from the latent space. Of course, a good VAE can improve the image quality by adding high-frequency details.

The VAE used in SDXL is a retrained one, using the same autoencoder architecture but with an increased batch size (256 versus 9) and, additionally, tracking the weights with an exponential moving average [2]. The new VAE outperforms the original model in all evaluated metrics.

Because of these implementation differences, instead of reusing the VAE code introduced in *Chapter 5*, we will need to write new code if we decide to use VAE independently. Here, we will provide an example of some common usage of the SDXL VAE:

1. Initialize a VAE model:

```
import torch
from diffusers.models import AutoencoderKL
vae_model = AutoencoderKL.from_pretrained(
    "stabilityai/stable-diffusion-xl-base-1.0",
    subfolder = "vae"
).to("cuda:0")
```

2. Encode an image using the VAE model. Before executing the following code, replace the `cat.png` file with a validated accessible image path:

```
from diffusers.utils import load_image
from diffusers.image_processor import VaeImageProcessor
image = load_image("/path/to/cat.png")

image_processor = VaeImageProcessor()
prep_image = image_processor.preprocess(image)
prep_image = prep_image.to("cuda:0")

with torch.no_grad():
    image_latent = vae_model.encode(prep_image
        ).latent_dist.sample()

image_latent.shape
```

3. Decode an image from latent space:

```
with torch.no_grad():
    decode_image = vae_model.decode(
        image_latent,
        return_dict = False
    )[0]

image = image_processor.postprocess(image = decode_image)[0]
image
```

In the preceding code, you first encode an image to latent space. An image in the latent space is invisible to our eyes, but it captures the features of an image in the latent space (in other words, in a high-dimensional vector space). Then, the decode part of the code decodes the image in the latent space to pixel space. From the preceding code, we know what the core functionality of a VAE is.

You might be curious as to why knowledge about the VAE is necessary. It has numerous applications. For instance, it allows you to save the generated latent image in a database and decode it only when needed. This method can reduce image storage by up to 90% without much loss of information.

The UNet of SDXL

The UNet model is the backbone neural network of SDXL. The UNet in SDXL is almost three times larger than the previous Stable Diffusion models. SDXL's UNet is a 2.6 GB billion parameter neural network, while the Stable Diffusion V1.5's UNet has 860 million parameters. Although the current open source LLM model is much larger in terms of neural network size, SDXL's UNet is, so far, the largest among those open source Diffusion models at the time of writing (October 2023), which directly leads to higher VRAM demands. 8 GB of VRAM can meet most of the use cases when using SD V1.5. For SDXL, 15 GB of VRAM is commonly required; otherwise, we will need to reduce the image resolution.

Besides the model size expansion, SDXL rearranges its Transformer block's position, which is crucial for better and more precise natural language-to-image guidance.

Two text encoders in SDXL

One of the most significant changes in SDXL is the text encoder. SDXL uses two text encoders together, CLIP ViT-L [5] and OpenCLIP ViT-bigG (also named OpenCLIP G/14). Furthermore, SDXL uses pooled embeddings from OpenCLIP ViT-bigG.

CLIP ViT-L is one of the most widely used models from OpenAI, which is also the text encoder or embedding model used in Stable Diffusion V1.5. What is the OpenCLIP ViT-bigG model? OpenCLIP is an open source implementation of **CLIP (Contrastive Language-Image Pre-Training)**. OpenCLIP G/14 is the largest and best OpenClip model trained on the LAION-2B dataset [9], a 100 TB dataset containing 2 billion images. While the OpenAI CLIP model generates a 768-dimensional embedding vector, OpenClip G/14 outputs a 1,280-dimensional embedding. By concatenating the two embeddings (of the same length), a 2,048-dimensional embedding is output. This is much larger than the previous 768-dimensional embedding from Stable Diffusion v1.5.

To illustrate the text encoding process, let's take the sentence a running dog as input; the ordinary text tokenizer will first convert the sentence into tokens, as shown in the following code:

```
input_prompt = "a running dog"

from transformers import CLIPTokenizer,CLIPTextModel
import torch

# initialize tokenizer 1
clip_tokenizer = CLIPTokenizer.from_pretrained(
    "stabilityai/stable-diffusion-xl-base-1.0",
    subfolder = "tokenizer",
    dtype = torch.float16
)

input_tokens = clip_tokenizer(
    input_prompt,
    return_tensors="pt"
)["input_ids"]
print(input_tokens)

clip_tokenizer_2 = CLIPTokenizer.from_pretrained(
```

```
    "stabilityai/stable-diffusion-xl-base-1.0",
    subfolder = "tokenizer_2",
    dtype = torch.float16
)

input_tokens_2 = clip_tokenizer_2(
    input_prompt,
    return_tensors="pt"
)["input_ids"]

print(input_tokens_2)
```

The preceding code will return the following result:

```
tensor([[49406,    320,  2761,  1929, 49407]])
tensor([[49406,    320,  2761,  1929, 49407]])
```

In the preceding result, 49406 is the beginning token and 49407 is the end token.

Next, the following code uses the CLIP text encoder to convert the tokens into embedding vectors:

```
clip_text_encoder = CLIPTextModel.from_pretrained(
    "stabilityai/stable-diffusion-xl-base-1.0",
    subfolder = "text_encoder",
    torch_dtype =torch.float16
).to("cuda")

# encode token ids to embeddings
with torch.no_grad():
    prompt_embeds = clip_text_encoder(
        input_tokens.to("cuda")
    )[0]

print(prompt_embeds.shape)
```

The result embedding tensor includes five 768 dimension vectors:

```
torch.Size([1, 5, 768])
```

The previous code used OpenAI's CLIP to convert the prompt text to 768-dimension embeddings. The following code uses the OpenClip G/14 model to encode the tokens into five 1,280-dimension embeddings:

```
clip_text_encoder_2 = CLIPTextModel.from_pretrained(
    "stabilityai/stable-diffusion-xl-base-1.0",
    subfolder = "text_encoder_2",
    torch_dtype =torch.float16
```

```
).to("cuda")

# encode token ids to embeddings
with torch.no_grad():
    prompt_embeds_2 = clip_text_encoder_2(input_tokens.to("cuda"))[0]

print(prompt_embeds_2.shape)
```

The result embedding tensor includes five 1,280-dimension vectors:

```
torch.Size([1, 5, 1280])
```

Now, the next question is, what are **pooled embeddings**? Embedding pooling is the process of converting a sequence of tokens into one embedding vector. In other words, pooling embedding is a lossy compression of information.

Unlike the embedding process we used before, which encodes each token into an embedding vector, a pooled embedding is one vector that represents the whole input text. We can generate the pooled embedding from OpenClip using the following Python code:

```
from transformers import CLIPTextModelWithProjection
clip_text_encoder_2 = CLIPTextModelWithProjection.from_pretrained(
    "stabilityai/stable-diffusion-xl-base-1.0",
    subfolder = "text_encoder_2",
    torch_dtype =torch.float16
).to("cuda")

# encode token ids to embeddings
with torch.no_grad():
    pool_embed = clip_text_encoder_2(input_tokens.to("cuda"))[0]

print(pool_embed.shape)
```

The preceding code will return a torch.Size([1, 1280]) pooled embedding vector from the text encoder. The maximum token size for a pooled embedding is 77. In SDXL, the pooled embedding is provided to the UNet together with the token-level embedding from both CLIP and OpenCLIP, guiding the image generation.

Don't worry – you won't need to manually provide these embeddings before using SDXL. StableDiffusionXLPipeline from the Diffusers package does everything for us. All we need to do is provide the prompt and negative prompt text. We will provide the sample code in the *Using SDXL* section.

The two-stage design

Another design addition in SDXL is its refiner model. According to the SDXL paper [2], the refiner model is used to enhance an image by adding more details and making it better, especially during the last 10 steps.

The refiner model is just another image-to-image model that can help fix broken images and add more elements to the images generated by the base model.

Based on my observations, for community-shared checkpoint models, the refiner model may not be necessary.

Next, we are going to use SDXL for common use cases.

Using SDXL

We briefly covered loading the SDXL model in *Chapter 6* and SDXL ControlNet usage in *Chapter 13*. You can find the sample codes there. In this section, we will cover more common SDXL usages, including loading community-shared SDXL models and how to use the image-to-image pipeline to enhance the model, using SDXL with community-shared LoRA models, and the unlimited length prompt pipeline from Diffuser (provided by the author of this book).

Use SDXL community models

Just months after the release of SDXL, the open source community has released countless fine-tuned SDXL models based on the base model from Stability AI. We can find these models on Hugging Face and CIVITAI (https://civitai.com/), and the number keeps growing.

Here, let's load one model from HuggingFace, using the SDXL model ID:

```
import torch
from diffusers import StableDiffusionXLPipeline
base_pipe = StableDiffusionXLPipeline.from_pretrained(
    "RunDiffusion/RunDiffusion-XL-Beta",
    torch_dtype = torch.float16
)
base_pipe.watermark = None
```

Note that in the preceding code, base_pipe.watermark = None will remove the invisible watermark from the generated image.

Next, move the model to CUDA, generate an image, and then offload the model from CUDA:

```
prompt = "realistic photo of astronaut cat in fighter cockpit,
detailed, 8k"
```

```
sdxl_pipe.to("cuda")
image = sdxl_pipe(
    prompt = prompt,
    width = 768,
    height = 1024,
    generator = torch.Generator("cuda").manual_seed(1)
).images[0]

sdxl_pipe.to("cpu")
torch.cuda.empty_cache()
image
```

With just one line prompt and not needing to provide any negative prompt, SDXL generates an amazing image, as shown in *Figure 16.2:*

Figure 16.2: A cat pilot generated by SDXL

You may want to use the refiner model to enhance the image, but the refiner model doesn't make a significant difference. Instead, we will use the image-to-image pipeline with the same model data to upscale the image.

Using SDXL image-to-image to enhance an image

Let's first upscale the image to twice:

```
from diffusers.image_processor import VaeImageProcessor
img_processor = VaeImageProcessor()
```

```
# get the size of the image
(width, height) = image.size

# upscale image
image_x = img_processor.resize(
    image = image,
    width = int(width * 1.5),
    height = int(height * 1.5)
)
image_x
```

Then, start an image-to-image pipeline by reusing the model data from the previous text-to-image pipeline, saving the RAM and VRAM usage:

```
from diffusers import StableDiffusionXLImg2ImgPipeline
img2img_pipe = StableDiffusionXLImg2ImgPipeline(
    vae = sdxl_pipe.vae,
    text_encoder = sdxl_pipe.text_encoder,
    text_encoder_2 = sdxl_pipe.text_encoder_2,
    tokenizer = sdxl_pipe.tokenizer,
    tokenizer_2 = sdxl_pipe.tokenizer_2,
    unet = sdxl_pipe.unet,
    scheduler = sdxl_pipe.scheduler,
    add_watermarker = None
)
img2img_pipe.watermark = None
```

Now, it is time to call the pipeline to further enhance the image:

```
img2img_pipe.to("cuda")
refine_image_2x = img2img_pipe(
    image = image_x,
    prompt = prompt,
    strength = 0.3,
    num_inference_steps = 30,
    guidance_scale = 4.0
).images[0]

img2img_pipe.to("cpu")
torch.cuda.empty_cache()
refine_image_2x
```

Note that we set the strength to 0.3 to preserve most of the original input image information. We will get a new, better image, as shown in *Figure 16.3*:

Figure 16.3: The refined cat pilot image from an image-to-image pipeline

While you might not notice many differences in this book at first glance, upon closer inspection of the image on a computer monitor, you will discover numerous additional details.

Now, let's explore how to utilize LoRA with Diffusers. If you're unfamiliar with LoRA, I recommend turning back to *Chapter 8*, which delves into the usage of Stable Diffusion LoRA in greater detail, and *Chapter 21*, which provides comprehensive coverage of LoRA training.

Using SDXL LoRA models

Not long ago, it was impossible to load LoRA using Diffusers, not to mention loading multiple LoRA models into one pipeline. With the massive work that has been done by the Diffusers team and community contributors, we can now load multiple LoRA models into the SDXL pipeline with the LoRA scale number specified.

And its usage is also extremely simple. It takes just two lines of code to add one LoRA to the pipeline:

```
sdxl_pipe.load_lora_weights("path/to/lora.safetensors")
sdxl_pipe.fuse_lora(lora_scale = 0.5)
```

To add two LoRA models:

```
sdxl_pipe.load_lora_weights("path/to/lora1.safetensors")
sdxl_pipe.fuse_lora(lora_scale = 0.5)

sdxl_pipe.load_lora_weights("path/to/lora2.safetensors")
sdxl_pipe.fuse_lora(lora_scale = 0.5)
```

As we discussed in *Chapter 8*, there are two ways to use LoRA – one is merging with the backbone model weights, and the other is dynamic monkey patching. Here, for SDXL, the method is model merging, which means unloading a LoRA from the pipeline. To unload a LoRA model, we will need to load the LoRA again but with a negative `lora_scale`. For example, if we want to unload `lora2.safetensors` from the pipeline, we can use the following code to achieve it:

```
sdxl_pipe.load_lora_weights("path/to/lora2.safetensors")
sdxl_pipe.fuse_lora(lora_scale = -0.5)
```

Besides using `fuse_lora` to load a LoRA model, we can also use PEFT-integrated LoRA loading. The code is very similar to the one we just used, but we add one more parameter called `adapter_name`, like this:

```
sdxl_pipe.load_lora_weights("path/to/lora1.safetensors",
    adapter_name="lora1")
sdxl_pipe.load_lora_weights("path/to/lora2.safetensors", ,
    adapter_name="lora2")
```

We can adjust the LoRA scale dynamically with the following code:

```
sdxl_pipe.set_adapters(["lora1", "lora2"], adapter_weights=[0.5, 1.0])
```

And we can also disable LoRA as follows:

```
sdxl_pipe.disable_lora()
```

Alternatively, we can disable one of the two loaded LoRA models, like this:

```
sdxl_pipe.set_adapters(["lora1", "lora2"], adapter_weights=[0.0, 1.0])
```

In the preceding code, we disabled `lora1` while continuing to use `lora2`.

With proper LoRA management code, you can use SDXL with an unlimited number of LoRA models. Speaking of "unlimited," next, we will cover the "unlimited" length prompt for SDXL.

Using SDXL with an unlimited prompt

By default, SDXL, like previous versions, supports only a maximum of 77 tokens for one-time image generation. In *Chapter 10*, we delved deep into implementing a text embedding encoder that supports weighted prompts without length limitation. For SDXL, the idea is similar but more complex and a bit harder to implement; after all, there are now two text encoders.

I built a long-weighted SDXL pipeline, `lpw_stable_diffusion_xl`, which is merged with the official `Diffusers` package. In this section, I will introduce the usage of this pipeline to enable a long-weighted and unlimited pipeline.

Make sure you have updated your `Diffusers` package to the latest version with the following command:

```
pip install -U diffusers
```

Then, use the following code to use the pipeline:

```
from diffusers import DiffusionPipeline
import torch

pipe = DiffusionPipeline.from_pretrained(
    "RunDiffusion/RunDiffusion-XL-Beta",
    torch_dtype = torch.float16,
    use_safetensors = True,
    variant = "fp16",
    custom_pipeline = "lpw_stable_diffusion_xl",
)

prompt = """
glamour photography, (full body:1.5) photo of young man,
white blank background,
wear sweater, with scarf,
wear jean pant,
wear nike run shoes,
wear sun glass,
wear leather shoes,
holding a umbrella in hand
""" * 2

prompt = prompt + " a (cute cat:1.5) aside"

neg_prompt = """
(worst quality:1.5),(low quality:1.5), paint, cg, spots, bad hands,
three hands, noise, blur
"""
```

```
pipe.to("cuda")
image = pipe(
    prompt = prompt,
    negative_prompt = neg_prompt,
    width = 832,
    height = 1216,
    generator = torch.Generator("cuda").manual_seed(7)
).images[0]

pipe.to("cpu")
torch.cuda.empty_cache()
image
```

The preceding code uses `DiffusionPipeline` to load a custom pipeline, `lpw_stable_diffusion_xl`, contributed by an open source community member (i.e., me).

Note that in the code, the prompt is multiplied by 2, making it definitely longer than 77 tokens. At the end of the prompt, a `(cute cat:1.5)` aside is appended. If the pipeline supports prompts longer than 77 tokens, there should be a cat in the generated result.

The image generated from the preceding code is shown in *Figure 16.4*:

Figure 16.4: A man with a cat, generated using an unlimited
prompt length pipeline – lpw_stable_diffusion_xl

From the image, we can see that all elements in the prompt are reflected, and there is now a cute cat sitting alongside the man.

Summary

This chapter covers the newest and best Stable Diffusion model – SDXL. We first introduced the basics of SDXL and why it is powerful and efficient, and then we drilled down into each component of the newly released model, covering VAE, UNet, text encoders, and the new two-stage design.

We provided a sample code for each of the components to help you understand SDXL inside out. These code samples can also be used to leverage the power of the individual components. For example, we can use VAE to compress images and a text encoder to generate text embeddings for images.

In the second half of this chapter, we covered some common use cases of SDXL, such as loading community-shared checkpoint models, using the image-to-image pipeline to enhance and upscale images, and introducing a simple and effective solution to load multiple LoRA models into one pipeline. Finally, we provided an end-to-end solution to use unlimited length-weighted prompts for SDXL.

With the help of SDXL, we can generate amazing images with short prompts and achieve much better results.

In the next chapter, we are going to discuss how to write Stable Diffusion prompts and leverage LLM to help produce and enhance prompts automatically.

References

1. SDXL: `https://stability.ai/stable-diffusion`

2. SDXL: Improving Latent Diffusion Models for High-Resolution Image Synthesis: `https://arxiv.org/abs/2307.01952`

3. Stable Diffusion XL Diffusers: `https://huggingface.co/docs/diffusers/main/en/using-diffusers/sdxl`

4. CLIP from OpenAI: `https://openai.com/research/clip`

5. CLIP VIT Large model: `https://huggingface.co/openai/clip-vit-large-patch14`

6. REACHING 80% ZERO-SHOT ACCURACY WITH OPENCLIP: VIT-G/14 TRAINED ON LAION-2B: `https://laion.ai/blog/giant-openclip/`

7. CLIP-ViT-bigG-14-laion2B-39B-b160k: `https://huggingface.co/laion/CLIP-ViT-bigG-14-laion2B-39B-b160k`

8. OpenCLIP GitHub repository: `https://github.com/mlfoundations/open_clip`

9. LAION-5B: A NEW ERA OF OPEN LARGE-SCALE MULTI-MODAL DATASETS: `https://laion.ai/blog/laion-5b/`

17
Building Optimized Prompts for Stable Diffusion

In Stable Diffusion V1.5 (SD V1.5), crafting prompts to generate ideal images can be challenging. It is not uncommon to see impressive images emerge from complex and unusual word combinations. This is largely due to the language text encoder used in Stable Diffusion V1.5 – OpenAI's CLIP model. CLIP is trained using captioned images from the internet, many of which are tags rather than structured sentences.

When using SD v1.5, we must not only memorize a plethora of "magical" keywords but also combine these tagging words effectively. For SDXL, its dual-language encoders, CLIP and OpenCLIP, are much more advanced and intelligent than those in the previous SD v1.5. However, we still need to follow certain guidelines to write effective prompts.

In this chapter, we will cover the fundamental principles for creating dedicated prompts and then explore powerful **large language model (LLM)** techniques to help us generate prompts automatically. Here are the topics we are going to cover in this chapter:

- What makes a good prompt?
- Using LLM as the prompt generator

Let's begin.

What makes a good prompt?

Some say using Stable Diffusion is like being a magician, where tiny tricks and alterations make a huge difference. Writing good prompts for Stable Diffusion is essential for getting the most out of this powerful text-to-image AI model. Let me introduce some best practices that will make your prompts more effective.

In the long run, AI models will understand natural language better and better, but for now, let's put in a bit of extra effort to make our prompts work better.

In the code files associated with this chapter, you will find that Stable Diffusion v1.5 is much more sensitive to prompts, as different prompts will significantly impact the outcome's image quality. Meanwhile, Stable Diffusion XL is much improved and is not so sensitive to prompts. In other words, a short prompt description for Stable Diffusion XL will generate relatively stable-quality images.

You can also find the code that generates all images in the code repository that comes with this chapter.

Be clear and specific

The more specific you are with your prompts, the more accurate the images you get from Stable Diffusion will be.

Here's an original prompt:

```
A painting of cool sci-fi.
```

From Stable Diffusion V1.5, we might get images such as those shown in *Figure 17.1*:

Figure 17.1: Images generated using SD V1.5 from the prompt "A painting of cool sci-fi"

It gives us animated human faces with advanced devices, but it is far from the "sci-fi" concept we might want.

From Stable Diffusion XL, the "sci-fi" concept is much more enriched, as shown in *Figure 17.2*:

Figure 17.2: Images generated using SDXL from the prompt "A painting of cool sci-fi"

The paintings are indeed cool, but short prompts generate images that are either not what we want or less controlled.

Now let's rewrite the prompt, adding more specific elements:

```
A photorealistic painting of a futuristic cityscape with towering
skyscrapers, neon lights, and flying vehicles, Science Fiction Artwork
```

With the improved prompt, SD V1.5 gives a much more accurate result than the original one, as shown in *Figure 17.3*:

Figure 17.3: Images generated using SD V1.5 from a prompt with specific elements added

SDXL also improves its output, reflecting the given prompt, as shown in *Figure 17.4*:

Figure 17.4: Images generated using SDXL from a prompt with specific elements added

Unless you purposefully let Stable Diffusion make its own decision, a good prompt clearly defines the desired outcome, leaving little room for ambiguity. It should specify the subject, style, and any additional details that characterize the image you envision.

Be descriptive

Descriptively describe the subject. This is similar to the *clear and specific* rule; not only should it be specific, but the more input and details we provide to the SD model, the better the result we will get. This is particularly effective for generating portrait images.

Say we want to generate a female portrait with the following prompt:

```
A beautiful woman
```

Here are the results we get from SD V1.5:

Figure 17.5: Images generated using SD V1.5 from the prompt "A beautiful woman"

The image is good overall but lacks details and seems half-painted, half-photo. SDXL generates better images with this short prompt, as shown in *Figure 17.6*:

Figure 17.6: Images generated using SDXL from the prompt "A beautiful woman"

But the outcome is random: sometimes a full-body image, sometimes fully face-focused. To better control the result, let's improve the prompt as follows:

```
Masterpiece, A stunning realistic photo of a woman with long, flowing
brown hair, piercing emerald eyes, and a gentle smile, set against a
backdrop of vibrant autumn foliage.
```

With this prompt, SD V1.5 returns better and more consistent images, as shown in *Figure 17.7*:

Figure 17.7: Images generated using SD V1.5 from an enhanced descriptive prompt

Similarly, SDXL also provides images scoped by the prompt instead of generating wild, out-of-control images, as shown in *Figure 17.8*:

Figure 17.8: Images generated using SDXL from an enhanced descriptive prompt

Mention details such as clothing, accessories, facial features, and the surrounding environment; the more, the better. Descriptiveness is crucial for guiding Stable Diffusion toward the desired image. Use descriptive language to paint a vivid picture in the Stable Diffusion model's "mind."

Using consistent terminology

Make sure the prompt is consistent throughout the context. Contradictory terminology will output unexpected results unless you are willing to be surprised by Stable Diffusion.

Say we give the following prompt, wanting to generate a man wearing a blue suit, but we also give `colorful cloth` as part of the keywords:

```
A man wears blue suit, he wears colorful cloth
```

This description is contradictory, and the SD model will be confused about what to generate: a blue suit or a colorful suit? The result is unknown. With this prompt, SDXL generated the two images shown in *Figure 17.9*:

Figure 17.9: Images generated using SD V1.5 from the prompt
"A man wears blue suit, he wears colorful cloth"

One image with a blue suit, another with a colorful suit. Let's improve the prompt to tell Stable Diffusion that we want a blue suit with a colorful scarf:

```
A man in a sharp, tailored blue suit is adorned with a vibrant,
colorful scarf, adding a touch of personality and flair to his
professional attire
```

Now, the result is much better and more consistent, as shown in *Figure 17.10*:

Figure 17.10: Images generated using SD V1.5 from a refined consistent prompt

Maintain consistency in the terminology you use to avoid confusing the model. If you refer to a key concept in the first part of a prompt, don't suddenly change to another in the latter part.

Reference artworks and styles

Reference specific artworks or artistic styles to guide the AI in replicating the desired aesthetic. Mention notable characteristics of the style, such as brushstrokes, color palettes, or compositional elements, which will heavily impact the generated results.

Let's generate an image of a night sky without mentioning Van Gogh's *Starry Night*:

```
A vibrant, swirling painting of a starry night sky with a crescent
moon illuminating a quaint village nestled among rolling hills."
```

Stable Diffusion V1.5 generates images with a cartoonish style, as shown in *Figure 17.11*:

Figure 17.11: Images generated using SD V1.5 from a prompt
without specifying a style or reference artwork

Let's add Van Gogh's Starry Night to the prompt:

```
A vibrant, swirling painting of a starry night sky reminiscent of
Van Gogh's Starry Night, with a crescent moon illuminating a quaint
village nestled among rolling hills.
```

The swirling style from Van Gogh is more dominant in the painting, as shown in *Figure 17.12*:

Figure 17.12: Images generated using SD V1.5 from a prompt with a style and reference artwork specified

Incorporate negative prompts

Stable Diffusion also provides a negative prompt input so that we can define elements that we don't want to be added to the image. Negative prompts function well in many situations.

We will use the following prompt, without applying a negative prompt:

```
1 girl, cute, adorable, lovely
```

Stable Diffusion will generate images as shown in *Figure 17.13*:

Figure 17.13: Images generated using SD V1.5 from a prompt without a negative prompt

This is not that bad, but far from good. Let's say we provide some negative prompts as follows:

```
paintings, sketches, worst quality, low quality, normal quality,
lowres,
monochrome, grayscale, skin spots, acne, skin blemishes, age spots,
extra fingers,
fewer fingers,broken fingers
```

The generated images are much improved, as shown in *Figure 17.14*:

Figure 17.14: Images generated using SD V1.5 from a prompt with negative prompts

Positive prompts add more attention to the target object for the Stable Diffusion model's UNet, while negative prompts take away the "attention" of the object displayed. Sometimes, simply adding appropriate negatives can greatly improve image quality.

Iterate and refine

Don't be afraid to experiment with different prompts and see what works best. It often takes some trial and error to get the perfect result.

However, manually creating prompts that meet these requirements is hard, not to mention a prompt that includes a subject, style, artist, resolution, details, color, and lighting information.

Next, we are going to employ an LLM as a prompt-generation helper.

Using LLMs to generate better prompts

All of the preceding rules or tips are helpful for a better understanding of how Stable Diffusion works with prompts. Since this is a book about using Stable Diffusion with Python, we don't want to handle these tasks by ourselves manually; the ultimate goal is to automate the whole process.

Stable Diffusion is evolving fast, and its cousins, the LLM and multi-modality community, are no slower. In this section, we are going to leverage LLMs to help us generate prompts with some keyword input. The following prompt will work for various kinds of LLM: ChatGPT, GPT-4, Google Bard, or any other capable open source LLM.

First, let's tell the LLM what it is going to do:

```
You will take a given subject or input keywords, and output a more
creative, specific, descriptive, and enhanced version of the idea in
the form of a fully working Stable Diffusion prompt. You will make all
prompts advanced, and highly enhanced. Prompts you output will always
have two parts, the "Positive Prompt" and "Negative prompt".
```

With the preceding prompt, the LLM knows what to do with the input; next, let's teach it a bit about Stable Diffusion. Without that, the LLM might have no idea what Stable Diffusion is:

```
Here is the Stable Diffusion document you need to know:

* Good prompts needs to be clear and specific, detailed and
descriptive.
* Good prompts are always consistent from beginning to end, no
contradictory terminology is included.
* Good prompts reference to artworks and style keywords, you are art
and style experts, and know how to add artwork and style names to the
prompt.

IMPORTANT:You will look through a list of keyword categories and
decide whether you want to use any of them. You must never use these
keyword category names as keywords in the prompt itself as literal
keywords at all, so always omit the keywords categories listed below:
     Subject
     Medium
     Style
     Artist
     Website
     Resolution
     Additional details
     Color
     Lighting
Treat the above keywords as a checklist to remind you what could be
used and what would best serve to make the best image possible.
```

We also need to tell the LLM the definitions of some terms:

```
About each of these keyword categories so you can understand them
better:

(Subject:)
The subject is what you want to see in the image.
(Resolution:)
The Resolution represents how sharp and detailed the image is. Let's
add keywords with highly detailed and sharp focus.
(Additional details:)
Any Additional details are sweeteners added to modify an image, such
as sci-fi, stunningly beautiful and dystopian to add some vibe to the
image.
(Color:)
color keywords can be used to control the overall color of the image.
The colors you specified may appear as a tone or in objects, such as
metallic, golden, red hue, etc.
(Lighting:)
Lighting is a key factor in creating successful images (especially
in photography). Lighting keywords can have a huge effect on how the
image looks, such as cinematic lighting or dark to the prompt.
(Medium:)
The Medium is the material used to make artwork. Some examples are
illustration, oil painting, 3D rendering, and photography.
(Style:)
The style refers to the artistic style of the image. Examples include
impressionist, surrealist, pop art, etc.
(Artist:)
Artist names are strong modifiers. They allow you to dial in the exact
style using a particular artist as a reference. It is also common to
use multiple artist names to blend their styles, for example, Stanley
Artgerm Lau, a superhero comic artist, and Alphonse Mucha, a portrait
painter in the 19th century could be used for an image, by adding this
to the end of the prompt:
by Stanley Artgerm Lau and Alphonse Mucha
(Website:)
The Website could be Niche graphic websites such as Artstation and
Deviant Art, or any other website which aggregates many images of
distinct genres. Using them in a prompt is a sure way to steer the
image toward these styles.
```

With the preceding definitions, we are teaching the LLM about the guidelines from the *What makes a good prompt?* section:

```
CRITICAL IMPORTANT: Your final prompt will not mention the category
names at all, but will be formatted entirely with these articles
omitted (A', 'the', 'there',) do not use the word 'no' in the Negative
prompt area. Never respond with the text, "The image is a", or "by
```

```
artist", just use "by [actual artist name]" in the last example
replacing [actual artist name] with the actual artist name when it's
an artist and not a photograph style image.

For any images that are using the medium of Anime, you will always
use these literal keywords at the start of the prompt as the first
keywords (include the parenthesis):

"masterpiece, best quality, (Anime:1.4)"

For any images that are using the medium of photo, photograph, or
photorealistic, you will always use all of the following literal
keywords at the start of the prompt as the first keywords (but  you
must omit the quotes):

"(((photographic, photo, photogenic))), extremely high quality high
detail RAW color photo"

Never include quote marks (this: ") in your response anywhere. Never
include, 'the image' or 'the image is' in the response anywhere.

Never include, too verbose of a sentence, for example, while being
sure to still share the important subject and keywords 'the overall
tone' in the response anywhere, if you have tonal keywords or keywords
just list them, for example, do not respond with, 'The overall tone of
the image is dark and moody', instead just use this:  'dark and moody'

The response you give will always only be all the keywords you have
chosen separated by a comma only.
```

Exclude any sexual or nude prompts:

```
IMPORTANT:
If the image includes any nudity at all, mention nude in the keywords
explicitly and do NOT provide these as keywords in the keyword prompt
area. You should always provide tasteful and respectful keywords.
```

Provide an example to the LLM as few-shot learning [1] material:

```
Here is an EXAMPLE (this is an example only):

I request: "A beautiful white sands beach"

You respond with this keyword prompt paragraph and Negative prompt
paragraph:

Positive Prompt: Serene white sands beach with crystal clear waters,
and lush green palm trees, Beach is secluded, with no crowds or
buildings, Small shells scattered across sand, Two seagulls flying
```

overhead. Water is calm and inviting, with small waves lapping at
shore, Palm trees provide shade, Soft, fluffy clouds in the sky, soft
and dreamy, with hues of pale blue, aqua, and white for water and
sky, and shades of green and brown for palm trees and sand, Digital
illustration, Realistic with a touch of fantasy, Highly detailed and
sharp focus, warm and golden lighting, with sun setting on horizon,
casting soft glow over the entire scene, by James Jean and Alphonse
Mucha, Artstation

Negative Prompt: low quality, people, man-made structures, trash,
debris, storm clouds, bad weather, harsh shadows, overexposure

Now, teach LLM how to output a negative prompt:

IMPORTANT: Negative Keyword prompts

Using negative keyword prompts is another great way to steer the
image, but instead of putting in what you want, you put in what you
don't want. They don't need to be objects. They can also be styles
and unwanted attributes. (e.g. ugly, deformed, low quality, etc.),
these negatives should be chosen to improve the overall quality of
the image, avoid bad quality, and make sense to avoid possible issues
based on the context of the image being generated, (considering its
setting and subject of the image being generated.), for example, if
the image is a person holding something, that means the hands will
likely be visible, so using 'poorly drawn hands' is wise in that case.

This is done by adding a 2nd paragraph, starting with the text
'Negative Prompt': and adding keywords. Here is a full example that
does not contain all possible options, but always use only what best
fits the image requested, as well as new negative keywords that would
best fit the image requested:

tiling, poorly drawn hands, poorly drawn feet, poorly drawn face, out
of frame, extra limbs, disfigured, deformed, body out of frame, bad
anatomy, watermark, signature, cut off, low contrast, underexposed,
overexposed, bad art, beginner, amateur, distorted face, blurry,
draft, grainy

IMPORTANT:
Negative keywords should always make sense in context to the image
subject and medium format of the image being requested. Don't add
any negative keywords to your response in the negative prompt keyword
area where it makes no contextual sense or contradicts, for example,
if I request: 'A vampire princess, anime image', then do NOT add
these keywords to the Negative prompt area: 'anime, scary, Man-made
structures, Trash, Debris, Storm clouds', and so forth. They need to
make sense of the actual image being requested so it makes sense in
context.

```
IMPORTANT:
For any images that feature a person or persons, and are also using
the Medium of a photo, photograph, or photorealistic in your response,
you must always respond with the following literal keywords at the
start of the NEGATIVE prompt paragraph, as the first keywords before
listing other negative keywords (omit the quotes):
"bad-hands-5, bad_prompt, unrealistic eyes"


If the image is using the Medium of an Anime, you must have these as
the first NEGATIVE keywords (include the parenthesis):
(worst quality, low quality:1.4)
```

Remind the LLM that there is a token limitation; here, you can change 150 to some other number.
The sample code associated with this chapter uses `lpw_stable_diffusion`, created by SkyTNT
[3], and `lpw_stable_diffusion_xl`, created by Andrew Zhu, the author of this book:

```
IMPORTANT: Prompt token limit:


The total prompt token limit (per prompt) is 150 tokens. Are you ready
for my first subject?
```

Put all prompts together in one chunk:

```
You will take a given subject or input keywords, and output a more
creative, specific, descriptive, and enhanced version of the idea in
the form of a fully working Stable Diffusion prompt. You will make all
prompts advanced, and highly enhanced. Prompts you output will always
have two parts, the "Positive Prompt" and "Negative prompt".


Here is the Stable Diffusion document you need to know:


* Good prompts needs to be clear and specific, detailed and
descriptive.
* Good prompts are always consistent from beginning to end, no
contradictory terminology is included.
* Good prompts reference to artworks and style keywords, you are art
and style experts, and know how to add artwork and style names to the
prompt.


IMPORTANT:You will look through a list of keyword categories and
decide whether you want to use any of them. You must never use these
keyword category names as keywords in the prompt itself as literal
keywords at all, so always omit the keywords categories listed below:
    Subject
    Medium
    Style
    Artist
    Website
```

```
Resolution
Additional details
Color
Lighting
```

About each of these keyword categories so you can understand them better:

(Subject:)

The subject is what you want to see in the image.

(Resolution:)

The Resolution represents how sharp and detailed the image is. Let's add keywords highly detailed and sharp focus.

(Additional details:)

Any Additional details are sweeteners added to modify an image, such as sci-fi, stunningly beautiful and dystopian to add some vibe to the image.

(Color:)

color keywords can be used to control the overall color of the image. The colors you specified may appear as a tone or in objects, such as metallic, golden, red hue, etc.

(Lighting:)

Lighting is a key factor in creating successful images (especially in photography). Lighting keywords can have a huge effect on how the image looks, such as cinematic lighting or dark to the prompt.

(Medium:)

The Medium is the material used to make artwork. Some examples are illustration, oil painting, 3D rendering, and photography.

(Style:)

The style refers to the artistic style of the image. Examples include impressionist, surrealist, pop art, etc.

(Artist:)

Artist names are strong modifiers. They allow you to dial in the exact style using a particular artist as a reference. It is also common to use multiple artist names to blend their styles, for example, Stanley Artgerm Lau, a superhero comic artist, and Alphonse Mucha, a portrait painter in the 19th century could be used for an image, by adding this to the end of the prompt:

by Stanley Artgerm Lau and Alphonse Mucha

(Website:)

The Website could be Niche graphic websites such as Artstation and Deviant Art, or any other website which aggregates many images of distinct genres. Using them in a prompt is a sure way to steer the image toward these styles.

Treat the above keywords as a checklist to remind you what could be used and what would best serve to make the best image possible.

CRITICAL IMPORTANT: Your final prompt will not mention the category names at all, but will be formatted entirely with these articles omitted (A', 'the', 'there',) do not use the word 'no' in the Negative prompt area. Never respond with the text, "The image is a", or "by artist", just use "by [actual artist name]" in the last example replacing [actual artist name] with the actual artist name when it's an artist and not a photograph style image.

For any images that are using the medium of Anime, you will always use these literal keywords at the start of the prompt as the first keywords (include the parenthesis):

"masterpiece, best quality, (Anime:1.4)"

For any images that are using the medium of photo, photograph, or photorealistic, you will always use all of the following literal keywords at the start of the prompt as the first keywords (but you must omit the quotes):

"(((photographic, photo, photogenic))), extremely high quality high detail RAW color photo"

Never include quote marks (this: ") in your response anywhere. Never include, 'the image' or 'the image is' in the response anywhere.

Never include, too verbose of a sentence, for example, while being sure to still share the important subject and keywords 'the overall tone' in the response anywhere, if you have tonal keywords or keywords just list them, for example, do not respond with, 'The overall tone of the image is dark and moody', instead just use this: 'dark and moody'

The response you give will always only be all the keywords you have chosen separated by a comma only.

IMPORTANT:
If the image includes any nudity at all, mention nude in the keywords explicitly and do NOT provide these as keywords in the keyword prompt area. You should always provide tasteful and respectful keywords.

Here is an EXAMPLE (this is an example only):

I request: "A beautiful white sands beach"

You respond with this keyword prompt paragraph and Negative prompt paragraph:

Positive Prompt: Serene white sands beach with crystal clear waters, and lush green palm trees, Beach is secluded, with no crowds or buildings, Small shells scattered across sand, Two seagulls flying overhead. Water is calm and inviting, with small waves lapping at shore, Palm trees provide shade, Soft, fluffy clouds in the sky, soft and dreamy, with hues of pale blue, aqua, and white for water and sky, and shades of green and brown for palm trees and sand, Digital illustration, Realistic with a touch of fantasy, Highly detailed and sharp focus, warm and golden lighting, with sun setting on horizon, casting soft glow over the entire scene, by James Jean and Alphonse Mucha, Artstation

Negative Prompt: low quality, people, man-made structures, trash, debris, storm clouds, bad weather, harsh shadows, overexposure

IMPORTANT: Negative Keyword prompts

Using negative keyword prompts is another great way to steer the image, but instead of putting in what you want, you put in what you don't want. They don't need to be objects. They can also be styles and unwanted attributes. (e.g. ugly, deformed, low quality, etc.), these negatives should be chosen to improve the overall quality of the image, avoid bad quality, and make sense to avoid possible issues based on the context of the image being generated, (considering its setting and subject of the image being generated.), for example, if the image is a person holding something, that means the hands will likely be visible, so using 'poorly drawn hands' is wise in that case.

This is done by adding a 2nd paragraph, starting with the text 'Negative Prompt': and adding keywords. Here is a full example that does not contain all possible options, but always use only what best fits the image requested, as well as new negative keywords that would best fit the image requested:

tiling, poorly drawn hands, poorly drawn feet, poorly drawn face, out of frame, extra limbs, disfigured, deformed, body out of frame, bad anatomy, watermark, signature, cut off, low contrast, underexposed, overexposed, bad art, beginner, amateur, distorted face, blurry, draft, grainy

IMPORTANT:
Negative keywords should always make sense in context to the image subject and medium format of the image being requested. Don't add any negative keywords to your response in the negative prompt keyword area where it makes no contextual sense or contradicts, for example, if I request: 'A vampire princess, anime image', then do NOT add these keywords to the Negative prompt area: 'anime, scary, Man-made structures, Trash, Debris, Storm clouds', and so forth. They need to make sense of the actual image being requested so it makes sense in context.

```
IMPORTANT:
For any images that feature a person or persons, and are also using
the Medium of a photo, photograph, or photorealistic in your response,
you must always respond with the following literal keywords at the
start of the NEGATIVE prompt paragraph, as the first keywords before
listing other negative keywords (omit the quotes):
"bad-hands-5, bad_prompt, unrealistic eyes"

If the image is using the Medium of an Anime, you must have these as
the first NEGATIVE keywords (include the parenthesis):
(worst quality, low quality:1.4)

IMPORTANT: Prompt token limit:

The total prompt token limit (per prompt) is 150 tokens. Are you ready
for my first subject?
```

Simply copy and paste the preceding prompt into ChatGPT, Bard, or any other serviceable LLM. Then, input any subject you want Stable Diffusion to generate:

```
a realistic photo of a woman standing by the side of a river
```

The LLM (Google Bard in this case) will provide us with a new prompt:

```
Positive Prompt:


(((photographic, photo, photogenic))), extremely high quality high
detail RAW color photo, A serene portrait of a woman standing by
the edge of a tranquil river, her gaze lost in the gentle flow of
the water. Her sun-kissed skin glows against the backdrop of lush
greenery, while her flowing hair cascades down her shoulders like
a waterfall. The warm sunlight filters through the trees, casting
dappled patterns on her face and the riverbank. The air is filled with
the sweet scent of wildflowers, and the gentle sound of birdsong fills
the air. The woman's expression is one of peace and tranquility, as if
she is completely at one with nature.


Negative Prompt:


bad-hands-5, bad_prompt, unrealistic eyes, blurry, low-resolution,
pixelated, noisy, grainy, overexposed, underexposed, harsh shadows,
bad lighting, out of focus, poorly composed, distracting elements, low
contrast, washed out, unrealistic anatomy, distorted face,
```

Using the original prompt a realistic photo of a woman standing by the side of a river, Stable Diffusion V1.5 generated the images shown in *Figure 17.15*:

Figure 17.15: Images generated using SD V1.5 from the original prompt
"a realistic photo of a woman standing by the side of a river"

With the new positive and negative prompts generated by the LLM, SD V1.5 generated the images shown in *Figure 17.16*:

Figure 17.16: Images generated using SD V1.5 from an LLM-generated prompt

The improvements apply to SDXL too. With the original prompt, SDXL generated the images shown in *Figure 17.17*:

Figure 17.17: Images generated using SDXL from the original prompt
"a realistic photo of a woman standing by the side of a river"

Using the LLM-generated positive and negative prompts, SDXL generated the images shown in *Figure 17.18*:

Figure 17.18: Images generated using SDXL from LLM-generated prompts

The images are undoubtedly better than the ones from the original prompt and prove that LLM-generated prompts can improve the quality of generated images.

Summary

In this chapter, we first discussed the challenges of composing prompts for Stable Diffusion to generate high-quality images. We then covered some fundamental rules for writing effective prompts for Stable Diffusion.

Taking it a step further, we summarized the rules of prompt writing and incorporated them into an LLM prompt. This approach not only works with ChatGPT [4] but also with other LLMs.

With the help of predefined prompts and LLMs, we can fully automate the image-generation process. There's no need to carefully write and tune prompts manually; simply ask the AI what you want to generate, and the LLM will provide sophisticated prompts and negative prompts. If set up correctly, Stable Diffusion can automatically execute the prompt and deliver the result without any human intervention.

We understand that the development speed of AI is rapid. In the near future, you will be able to add more of your own LLM prompts to make the process even smarter and more powerful. This will further enhance the capabilities of Stable Diffusion and LLMs, allowing you to generate stunning images with minimal effort.

In the next chapter, we'll use the knowledge we learned from the previous chapters to build useful applications using Stable Diffusion.

References

1. *Language Models are Few-Shot Learners*: https://arxiv.org/abs/2005.14165

2. *Best text prompt for creating Stable diffusion prompts through ChatGPT or a local LLM model? What do you use that is better?*: https://www.reddit.com/r/StableDiffusion/comments/14to15n/best_text_prompt_for_creating_stable_diffusion/

3. SkyTNT: https://github.com/SkyTNT?tab=repositories

4. ChatGPT: https://chat.openai.com/

Part 4 – Building Stable Diffusion into an Application

Throughout this book, we've explored the vast potential of Stable Diffusion, from its fundamental concepts to advanced applications and customization techniques. Now, it's time to bring everything together and integrate Stable Diffusion into real-world applications, making its power accessible to users and unlocking new possibilities for creative expression and problem-solving.

In this final part, we'll focus on building practical applications that showcase the versatility and impact of Stable Diffusion. You'll learn how to develop innovative solutions such as object editing and style transferring, enabling users to manipulate images in unprecedented ways. We'll also cover the importance of data persistence, demonstrating how to save image generation prompts and parameters directly within the generated PNG images.

Furthermore, you'll discover how to create interactive user interfaces using popular frameworks such as Gradio, making it easy for users to engage with Stable Diffusion models. Additionally, we'll delve into the realm of transfer learning, guiding you through the process of training a Stable Diffusion LoRA from scratch. Finally, we'll conclude with a broader discussion on the future of Stable Diffusion, AI, and the importance of staying informed about the latest developments in this rapidly evolving field.

By the end of this part, you'll be equipped with the knowledge and skills necessary to integrate Stable Diffusion into a wide range of applications, from creative tools to productivity-enhancing software. The possibilities are endless, and it's time to unleash the full potential of Stable Diffusion!

This part contains the following chapters:

- *Chapter 18, Applications – Object Editing and Style Transferring*
- *Chapter 19, Generation Data Persistence*
- *Chapter 20, Creating Interactive User Interfaces*
- *Chapter 21, Diffusion Model Transfer Learning*
- *Chapter 22, Exploring Beyond Stable Diffusion*

18

Applications – Object Editing and Style Transferring

Stable Diffusion (SD) is not only capable of generating a variety of images but it can also be utilized for image editing and style transfer from one image to another. In this chapter, we will explore solutions for image editing and style transfer.

Along the way, we will also introduce the tools that enable us to achieve these goals: **CLIPSeg**, which is used to detect the content of an image; **Rembg**, which is a tool that flawlessly removes the background of an image; and **IP-Adapter**, which is used to transfer the style from one image to another.

In this chapter, we are going to cover the following topics:

- Editing images using Stable Diffusion
- Object and style transferring

Let's start.

Editing images using Stable Diffusion

Do you recall the background swap example we discussed in *Chapter 1*? In this section, we will introduce a solution that can assist you in editing the content of an image.

Before we can edit anything, we need to identify the boundary of the object we want to edit. In our case, to obtain the background mask, we will use the CLIPSeg [1] model. **CLIPSeg**, which stands for **CLIP-based Image Segmentation**, is a model trained to segment images based on text prompts or reference images. Unlike traditional segmentation models that require a large amount of labeled data, CLIPSeg can achieve impressive results with little to no training data.

CLIPSeg builds upon the success of CLIP, the same model used by SD. CLIP is a powerful pre-trained model that learns to connect text and images. The CLIPSeg model adds a small decoder module on top of CLIP, allowing it to translate the learned relationships into pixel-level segmentation. This means we can provide CLIPSeg with a simple description such as "the background of this picture," and CLIPSeg will return the mask of the targeted objects.

Now, let's see how we can use CLIPSeg to accomplish some tasks.

Replacing image background content

We will first load up the CLIPSeg processor and model, then provide both the prompt and image to the model to generate the mask data, and finally, use the SD inpainting pipeline to redraw the background. Let's do it step by step:

1. Load the CLIPSeg model.

   The following code will load up the CLIPSegProcessor processor and CLIPSegForImageSegmentation model:

   ```
   from transformers import(
       CLIPSegProcessor,CLIPSegForImageSegmentation)

   processor = CLIPSegProcessor.from_pretrained(
       "CIDAS/clipseg-rd64-refined"
   )
   model = CLIPSegForImageSegmentation.from_pretrained(
       "CIDAS/clipseg-rd64-refined"
   )
   ```

   The processor will be used to preprocess both the prompt and images input. The model will be the one responsible for model inference.

2. Generate the grayscale mask.

   By default, the CLIPSeg model will return logits of its result. By applying the torch.sigmoid() function, we can then have the grayscale mask of the target object in the image. The grayscale mask can then enable us to generate the binary mask, which will be used in the SD inpainting pipeline:

   ```
   from diffusers.utils import load_image
   from diffusers.utils.pil_utils import numpy_to_pil
   import torch

   source_image = load_image("./images/clipseg_source_image.png")

   prompts = ['the background']
   inputs = processor(
       text = prompts,
   ```

```
        images = [source_image] * len(prompts),
        padding = True,
        return_tensors = "pt"
    )

    with torch.no_grad():
        outputs = model(**inputs)

    preds = outputs.logits
    mask_data = torch.sigmoid(preds)

    mask_data_numpy = mask_data.detach().unsqueeze(-1).numpy()
    mask_pil = numpy_to_pil(
        mask_data_numpy)[0].resize(source_image.size)
```

The preceding code will generate a grayscale mask image that highlights the background, as shown in *Figure 18.1*:

Figure 18.1: Background grayscale mask

This mask is still not the one we want; we need a binary mask. Why do we need a binary mask? Because SD v1.5 inpainting works better with a binary mask than a grayscale mask. You may also add the grayscale mask to the SD pipeline to see the result; there's nothing to lose by trying different combinations and inputs.

3. Generate a binary mask.

    We will use the following code to convert a grayscale mask into a 0-1 binary mask image:

    ```
    bw_thresh = 100
    bw_fn = lambda x : 255 if x > bw_thresh else 0
    bw_mask_pil = mask_pil.convert("L").point(bw_fn, mode="1")
    ```

Let me explain the key elements we presented in the preceding code:

- `bw_thresh`: This defines the threshold of treating a pixel as black or white. In the preceding code, any grayscale pixel value higher than 100 will be treated as a white highlight.

- `mask_pil.convert("L")`: This converts the `mask_pil` image into grayscale mode. Grayscale images have only one channel, representing pixel intensity values from 0 (black) to 255 (white).

- `.point(bw_fn, mode="1")`: This applies the `bw_fn` thresholding function to each pixel of the grayscale image. The `mode="1"` argument ensures that the output image is a 1-bit binary image (black and white only).

We will see the result shown in *Figure 18.2*:

Figure 18.2: Background binary mask

4. Redraw the background using the SD inpainting model:

```
from diffusers import(StableDiffusionInpaintPipeline,
    EulerDiscreteScheduler)
inpaint_pipe = StableDiffusionInpaintPipeline.from_pretrained(
    "CompVis/stable-diffusion-v1-4",
    torch_dtype = torch.float16,
    safety_checker = None
).to("cuda:0")

sd_prompt = "blue sky and mountains"
out_image = inpaint_pipe(
```

```
        prompt = sd_prompt,
        image = source_image,
        mask_image = bw_mask_pil,
        strength = 0.9,
        generator = torch.Generator("cuda:0").manual_seed(7)
    ).images[0]
    out_image
```

In the preceding code, we use the SD v1.4 model as the inpainting model because it generates better results than the SD v1.5 model. If you execute it, you will see the exact result we presented in *Chapter 1*. The background is now no longer a vast planetary universe but blue sky and mountains.

The same technique can be used for many other purposes, such as editing clothing in a photo and adding items to a photo.

Removing the image background

Many times, we want to just remove the background of an image. With the binary mask in hand, removing the background isn't hard at all. We can do it using the following code:

```
from PIL import Image, ImageOps
output_image = Image.new("RGBA", source_image.size,
    (255,255,255,255))
inverse_bw_mask_pil = ImageOps.invert(bw_mask_pil)
r = Image.composite(source_image ,output_image,
    inverse_bw_mask_pil)
```

Here's a breakdown of what each line does:

- `from PIL import Image, ImageOps`: This line imports the `Image` and `ImageOps` modules from PIL. The `Image` module provides a class with the same name that is used to represent a PIL image. The `ImageOps` module contains a number of "ready-made" image-processing operations.

- `output_image = Image.new("RGBA", source_image.size, (255,255,255,255))`: This line creates a new image with the same size as `source_image`. The new image will be in RGBA mode, meaning it includes channels for red, green, blue, and alpha (transparency). The initial color of all pixels in the image is set to white (255,255,255) with full opacity (255).

- `inverse_bw_mask_pil = ImageOps.invert(bw_mask_pil)`: This line inverts the colors of the `bw_mask_pil` image using the `invert` function from ImageOps. If `bw_mask_pil` is a black and white image, the result will be a negative of the original image, that is, black becomes white and white becomes black.

- `r = Image.composite(source_image ,output_image, inverse_bw_mask_ pil)`: This line creates a new image by blending `source_image` and `output_image` based on the `inverse_bw_mask_pil` mask image. Where the mask image is white (or shades of gray), the corresponding pixels from `source_image` are used, and where the mask image is black, the corresponding pixels from `output_image` are used. The result is assigned to `r`.

Simply four lines of code enable the replacement of the background with pure white, as shown in *Figure 18.3*:

Figure 18.3: Remove background using CLIPSeg

But, we will see jagged edges; this is not good and can't be perfectly solved using CLIPSeg. If you are going to feed this image into the diffusion pipeline again, SD will help fix the jagged edges problem by using another image-to-image pipeline. Based on the nature of the diffusion model, the background edges will be either blurred or rerendered with other pixels. To remove the background neatly, we will need other tools to help, for example, the Rembg project [2]. Its usage is also simple:

1. Install the package:

    ```
    pip install rembg
    ```

2. Remove the background with two lines of code:

    ```
    from rembg import remove
    remove(source_image)
    ```

And we see the background is completely removed, as shown in *Figure 18.4*:

Figure 18.4: Remove background using Rembg

To set the background as white, use three more lines of code, as shown below:

```
from rembg import remove
from PIL import Image
white_bg = Image.new("RGBA", source_image.size, (255,255,255))
image_wo_bg = remove(source_image)
Image.alpha_composite(white_bg, image_wo_bg)
```

We can find the background is completely replaced with a white background. An object with a pure white background can be useful in some cases; for instance, we are going to use the object as a guidance embedding. No, you did not read that wrong; we can use the image as the input prompt. Let's explore this in the next section.

Object and style transferring

When we introduced the theory behind SD in *Chapters 4* and *5*, we learned that only text embedding is involved in the UNet diffusion process. Even if we provide an initial image as the starting point, the initial image is simply used as the starting noise or concatenated with initial noises. It does not have any influence on the steps of the diffusion process.

That is until the IP-Adapter project [3] came about. IP-Adapter is a tool that lets you use an existing image as a reference for text prompts. In other words, we can take the image as another piece of prompt work together with text guidance to generate an image. Unlike Textual Inversion, which usually works well for certain concepts or styles, IP-Adapter works with any images.

With the help of IP-Adapter, we can magically transfer an object from one image to a completely different one.

Next, let's start using IP-Adapter to transfer an object from one image to another one.

Loading up a Stable Diffusion pipeline with IP-Adapter

Using IP-Adapter in Diffusers is simple enough, you don't need to install any additional packages or manually download any model files:

1. Load the image encoder. It is this dedicated image encoder that plays a key role in turning the image into a guidance prompt embedding:

```
import torch
from transformers import CLIPVisionModelWithProjection
image_encoder = CLIPVisionModelWithProjection.from_pretrained(
    "h94/IP-Adapter",
    subfolder = "models/image_encoder",
    torch_dtype = torch.float16,
).to("cuda:0")
```

2. Load a vanilla SD pipeline but with one additional `image_encoder` parameter:

```
from diffusers import StableDiffusionImg2ImgPipeline
pipeline = StableDiffusionImg2ImgPipeline.from_pretrained(
    "runwayml/stable-diffusion-v1-5",
    image_encoder = image_encoder,
    torch_dtype = torch.float16,
    safety_checker = None
).to("cuda:0")
```

> **Note**
>
> We will use the image encoder model from `models/image_encoder` even when loading an SDXL pipeline, rather than `sdxl_models/image_encoder`; otherwise, an error message will be thrown. You can also replace the SD v1.5 base model with any other community-shared models.

3. Apply IP-Adapter to the UNet pipeline:

```
pipeline.load_ip_adapter(
    "h94/IP-Adapter",
```

```
        Subfolder = "models",
        weight_name = "ip-adapter_sd15.bin"
    )
```

If you are using an SDXL pipeline, replace `models` with `sdxl_models`, and replace `ip-adapter_sd15.bin` with `ip-adapter_sdxl.bin`.

That is all; now we can use the pipeline just like any other pipeline. Diffusers will help you download the model files automatically if no IP-Adapter models exist. In the next section, we are going to use the IP-Adapter model to transfer a style from one image to another.

Transferring style

In this section, we are going to write code to transfer the famous *Girl with a Pearl Earring* by Johannes Vermeer (see *Figure 18.5*) to the *astronaut riding a horse* image:

Figure 18.5: Girl with a Pearl Earring by Johannes Vermeer

Here, let's kick off the pipeline to transfer style:

```
from diffusers.utils import load_image

source_image = load_image("./images/clipseg_source_image.png")
ip_image = load_image("./images/vermeer.png")

pipeline.to("cuda:0")

image = pipeline(
```

```
    prompt = 'best quality, high quality',
    negative_prompt = "monochrome,lowres, bad anatomy,low quality" ,
    image = source_image,
    ip_adapter_image = ip_image ,
    num_images_per_prompt = 1 ,
    num_inference_steps = 50,
    strength = 0.5,
    generator = torch.Generator("cuda:0").manual_seed(1)
).images[0]

pipeline.to("cpu")
torch.cuda.empty_cache()
image
```

In the preceding code, we used the original astronaut image – source_image – as the base, and the oil painting image as the IP-Adapter image prompt – ip_image (we want its style). Amazingly, we get the result shown in *Figure 18.6*:

Figure 18.6: Astronaut riding a horse with a new style

The style and feel of the *Girl with a Pearl Earring* image have successfully been applied to another image.

IP-Adapter's potential is huge. We can even transfer the clothing and face from one image to another. More usage samples can be found in the original IP-Adapter repository [3] and the Diffusers PR page [5].

Summary

In this chapter, the focus was on using SD for image editing and style transferring. The chapter introduced tools such as CLIPSeg for image content detection, Rembg for background removal, and IP-Adapter for transferring styles between images.

The first section covered image editing, specifically replacing or removing the background. CLIPSeg is used to generate a mask of the background, which is then converted to a binary mask. The background is either replaced using SD or removed, with the latter option showing jagged edges. Rembg was introduced as a solution for smoother background removal.

The second section explored object and style transferring using IP-Adapter. The process involves loading an image encoder, incorporating it into an SD pipeline, and applying IP-Adapter to the UNet of the pipeline. The chapter concluded with an example of transferring the style of Vermeer's *Girl with a Pearl Earring* onto an image of an astronaut riding a horse.

In the next chapter, we are going to explore solutions to save and read the parameters and prompt information to and from the generated image files.

References

1. CLIPSeg GitHub repository: `https://github.com/timojl/clipseg`

2. Timo Lüddecke and Alexander S. Ecker, *Image Segmentation Using Text and Image Prompts*: `https://arxiv.org/abs/2112.10003`

3. IP-Adapter GitHub repository: `https://github.com/tencent-ailab/IP-Adapter`

4. Rembg, a tool to remove image backgrounds: `https://github.com/danielgatis/rembg`

5. IP-Adapters original samples: `https://github.com/huggingface/diffusers/pull/5713`

Generation Data Persistence

Imagine a Python program generates images but when you go back to the image hoping to make improvements or simply generate new images based on the original prompt, you can't find the exact prompt, inference steps, guidance scale, and the other things that actually generate the image!

One of the solutions to solve this problem is saving all the metadata in the generated image file. The **Portable Network Graphics (PNG)** [1] image format provides a mechanism for us to store a piece of metadata along with the image pixel data. We will explore this solution.

In this chapter, we are going to look at the following:

- Exploring and understanding the PNG file structure
- Storing the Stable Diffusion generation metadata in the PNG file
- Extracting the Stable Diffusion generation metadata from the PNG file

By employing the solution provided by this chapter, you will be able to maintain the generation prompt and parameters in the image file, and also extract the meta information for further usage.

Let's start.

Exploring and understanding the PNG file structure

Before saving the image metadata and Stable Diffusion generation parameters in the image, we'd better have an overall understanding of why we're choosing PNG as the output image format to save the output of Stable Diffusion, and why PNG can support unlimited custom metadata, which is useful for writing a large amount of data into the image.

By understanding the PNG format, we can confidently write and read data into a PNG file as we are going to persist data in the image.

PNG is a raster graphics file format, an ideal image format for images generated by Stable Diffusion. The PNG file format was created as an improved, non-patented lossless image compression format, and is now widely used on the internet.

Besides PNG, several other image formats also support saving custom image metadata, such as JPEG, TIFF, RAW, DNG, and BMP. However, these formats have their problems and limitations. JPEG files can include custom Exif metadata, but JPEG is a loss compression image format, reaching its high rate of compression by sacrificing image quality. DNG is a proprietary format owned by Adobe. BMP's custom metadata size is limited compared with PNG.

For the PNG format, besides the capability of storing additional metadata, there are a lot of advantages that make it an ideal format [1]:

- **Lossless compression**: PNG uses lossless compression, which means the image quality is not degraded when compressed

- **Transparency support**: PNG supports transparency (alpha channel), allowing images to have transparent backgrounds or semi-transparent elements

- **Wide color range**: PNG supports 24-bit RGB color, 32-bit RGBA color, and grayscale images, providing a wide range of color options

- **Gamma correction**: PNG supports gamma correction, which helps maintain consistent colors across different devices and platforms

- **Progressive display**: PNG supports interlacing, allowing the image to be displayed progressively as it is being downloaded

We also need to be aware that, in some cases, PNG may not be the best choice. Here, let me list some:

- **Larger file size**: Compared to other formats such as JPEG, PNG files can be larger due to its lossless compression

- **No native support for animation**: Unlike GIF, PNG does not support animation natively

- **Not suitable for high-resolution photographs**: Due to its lossless compression, PNG is not the best choice for high-resolution photographs, as the file size can be significantly larger than formats such as JPEG that use lossy compression

Despite these limitations, PNG remains a viable option for image formatting, particularly for raw images from Stable Diffusion.

The internal data structure of a PNG file is based on a chunk-based structure. Each chunk is a self-contained unit that stores specific information about the image or metadata. This structure allows PNG files to store additional information, such as text, copyright, or any other metadata, without affecting the image data itself.

A PNG file consists of a signature followed by a series of chunks. Here's a brief overview of the main components of a PNG file:

- **Signature**: The first 8 bytes of a PNG file are a fixed signature (89 50 4E 47 0D 0A 1A 0A in hexadecimal) that identifies the file as a PNG.

- **Chunks**: The rest of the file is composed of chunks. Each chunk has the following structure:

  - **Length** (4 bytes): An unsigned integer representing the length of the chunk's data field in bytes.

  - **Type** (4 bytes): A 4-byte string that specifies the type of the chunk (e.g., IHDR, IDAT, tEXt, etc.).

  - **Data** (variable length): The chunk's data, as specified by the `length` field.

  - **CRC** (4 bytes): A **cyclic redundancy check** (**CRC**) value for error detection, calculated based on the chunk's type and data fields.

This structure offers both flexibility and extensibility, as it allows for the addition of new chunk types without disrupting the compatibility with existing PNG decoders. Moreover, this PNG data structure enables the insertion of nearly limitless additional metadata into the image.

Next, we will utilize Python to insert some text data into a PNG image file.

Saving extra text data in a PNG image file

First and foremost, let's use Stable Diffusion to generate an image for testing. Not like the code we used in previous chapters, this time, we are going to use a JSON object to store the generation parameters.

Load the model:

```python
import torch
from diffusers import StableDiffusionPipeline

model_id = "stablediffusionapi/deliberate-v2"
text2img_pipe = StableDiffusionPipeline.from_pretrained(
    model_id,
    torch_dtype = torch.float16
)
# Then, we define all the parameters that will be used to generate an
# image in a JSON object:
gen_meta = {
    "model_id": model_id,
    "prompt": "high resolution,
        a photograph of an astronaut riding a horse",
    "seed": 123,
    "inference_steps": 30,
    "height": 512,
    "width": 768,
    "guidance_scale": 7.5
}
```

Now, let's use `gen_meta` in the Python `dict` type:

```
text2img_pipe.to("cuda:0")
input_image = text2img_pipe(
    prompt = gen_meta["prompt"],
    generator = \
        torch.Generator("cuda:0").manual_seed(gen_meta["seed"]),
    guidance_scale = gen_meta["guidance_scale"],
    height = gen_meta["height"],
    width = gen_meta["width"]
).images[0]
text2img_pipe.to("cpu")
torch.cuda.empty_cache()
input_image
```

We should have an image generated with the `input_image` handle – the reference to the image object in the Python context.

Next, let's store the `gen_meta` data in the PNG file step by step:

1. Install the `pillow` library [2] if you haven't already:

    ```
    pip install pillow
    ```

2. Use the following code to add one chunk that stores text information:

    ```
    from PIL import Image
    from PIL import PngImagePlugin
    import json

    # Open the original image
    image = Image.open("input_image.png")

    # Define the metadata you want to add
    metadata = PngImagePlugin.PngInfo()
    gen_meta_str = json.dumps(gen_meta)
    metadata.add_text("my_sd_gen_meta", gen_meta_str)

    # Save the image with the added metadata
    image.save("output_image_with_metadata.png", "PNG",
        pnginfo=metadata)
    ```

    Now the stringified `gen_meta` is in the `output_image_with_metadata.png` file. Please note that we need to first convert `gen_data` from an object to a string using `json.dumps(gen_meta)`.

The preceding code added one chunk of data to the PNG file. As we learned at the beginning of this chapter, the PNG file is stacked in chunks, which means we should be able to add any number of text chunks to the PNG file. In the following example, we added two chunks instead of just one:

```
from PIL import Image
from PIL import PngImagePlugin
import json

# Open the original image
image = input_image#Image.open("input_image.png")

# Define the metadata you want to add
metadata = PngImagePlugin.PngInfo()
gen_meta_str = json.dumps(gen_meta)
metadata.add_text("my_sd_gen_meta", gen_meta_str)

# add a copy right json object
copyright_meta = {
    "author":"Andrew Zhu",
    "license":"free use"
}
copyright_meta_str = json.dumps(copyright_meta)
metadata.add_text("copy_right", copyright_meta_str)

# Save the image with the added metadata
image.save("output_image_with_metadata.png", "PNG",
    pnginfo=metadata)
```

Simply by calling another add_text() function, we can add a second text chunk to the PNG file. Next, let's extract the added data from the PNG image.

3. Extracting text data from a PNG image is straightforward. We will use the pillow package again for the extraction task:

```
from PIL import Image
image = Image.open("output_image_with_metadata.png")

metadata = image.info

# print the meta
for key, value in metadata.items():
    print(f"{key}: {value}")
```

We should see an output like this:

```
my_sd_gen_meta: {"model_id": "stablediffusionapi/deliberate-v2",
"prompt": "high resolution, a photograph of an astronaut riding
a horse", "seed": 123, "inference_steps": 30, "height": 512,
"width": 768, "guidance_scale": 7.5}
copy_right: {"author": "Andrew Zhu", "license": "free use"}
```

With the code provided in this section, we should be able to save and retrieve custom data to and from a PNG image file.

PNG extra data storage limitation

You may wonder whether there are any limitations on text data size. There is no specific limit on the amount of metadata that can be written to a PNG file. However, there are practical constraints based on the PNG file structure and the limitations of the software or libraries used to read and write the metadata.

A PNG file, as we discussed in the first section, is stored in chunks. Each chunk has a maximum size of $2^{31} - 1$ bytes (approximately 2 GB). While it is theoretically possible to include multiple metadata chunks within a single PNG file, storing excessive or overly large data within these chunks can lead to errors or slow loading times when attempting to open the image using other software.

In practice, metadata in PNG files is usually small, containing information such as copyright, author, description, or the software used to create the image. In our case, it is the Stable Diffusion parameters that are used to generate the image. It is not recommended to store large amounts of data in PNG metadata, as it may cause performance issues and compatibility problems with some software.

Summary

In this chapter, we introduced a solution to store the image generation prompt and relative parameters in the PNG image file, so that the generation data will go with the file wherever it goes and we can extract the parameters, using Stable Diffusion to enhance it or extend the prompt for other usage.

This chapter introduced the file structure of a PNG file and provided sample code to store multiple chunks of text data in a PNG file and then use Python code to extract the metadata from the PNG file.

With the solution's sample code, you will be able to extract the metadata from an image generated by A1111's Stable Diffusion web UI too.

In the next chapter, we will build an interactive web UI for a Stable Diffusion application.

References

1. Portable Network Graphics (PNG) specification: https://www.w3.org/TR/png/
2. Pillow package: https://pillow.readthedocs.io/en/stable/

20

Creating Interactive User Interfaces

In previous chapters, we used only Python code and Jupyter Notebook to achieve various tasks using Stable Diffusion. In some scenarios, we need an interactive user interface not only for easier testing but also for a better user experience.

Imagine we have built an application using Stable Diffusion. How can we publish it to the public or non-technical users to try it out? In this chapter, we are going to use an open sourced interactive UI framework, Gradio [1], to encapsulate diffusers code and provide a web-based UI, using only Python.

This chapter won't delve into every aspect of Gradio usage. Instead, we'll focus on giving you a high-level overview of its fundamental building blocks, all with a specific goal in mind: demonstrating how to construct a Stable Diffusion text-to-image pipeline using Gradio.

In this chapter, we will cover these topics:

- Introducing Gradio
- Gradio fundamentals
- Building a Stable Diffusion text-to-image pipeline with Gradio

Let's start.

Introducing Gradio

Gradio is a Python library that makes it easy to build beautiful, interactive web interfaces for machine learning models and data science workflows. It is a high-level library that abstracts away the details of web development so you can focus on building your model and interface.

The A1111 Stable Diffusion Web UI that we mentioned several times in previous chapters uses Gradio as the user interface, and many researchers use this framework for a quick demo of their most recent work. Here are some reasons why Gradio is the prevailing user interface:

- **Easy to use**: Gradio's simple API makes it easy to create interactive web interfaces in just a few lines of code
- **Flexible**: Gradio can be used to create a wide variety of interactive web interfaces, from simple sliders to complex chatbots
- **Extensible**: Gradio is extensible, so you can customize the look and feel of your interfaces or add new features
- **Open source**: Gradio is open source, so you can contribute to the project or use it in your projects

Another feature of Gradio that doesn't exist in other similar frameworks is that Gradio interfaces can be embedded in Python notebooks or presented as standalone web pages (you will find out why this notebook embedding feature is cool when you see it).

If you have been running Stable Diffusion using diffusers, your Python environment should be ready for Gradio. In case this is the first chapter of your reading journey, make sure you have Python 3.8 or a higher version installed on your machine.

Now that we know what Gradio is, let's learn how to set it up.

Getting started with Gradio

In this section, we will learn about the bare minimum setup needed to spin up a Gradio application:

1. Install Gradio using `pip`:

    ```
    pip install gradio
    ```

    Please also ensure you update the following two packages to the newest version: `click` and `uvicorn`:

    ```
    pip install -U click
    pip install -U uvicorn
    ```

2. Create a Jupyter Notebook cell and write or copy the following code in the cell:

```
import gradio

def greet(name):
    return "Hello " + name + "!"

demo = gradio.Interface(
    fn = greet,
    inputs = "text",
    outputs = "text"
)

demo.launch()
```

Executing it will not pop out a new web browser window. Instead, the UI will be embedded inside the Jupyter Notebook result panel.

Of course, you can copy and paste the local URL – `http://127.0.0.1:7860` – to any local browser to view it.

Be aware that the next time you execute the code in another Jupyter Notebook's cell, a new server port will be allocated, such as `7861`. Gradio won't take back the assigned server port automatically. We can use one additional line of code – `gr.close_all()` – to ensure all liveports are released before the launch. Update the code as shown here:

```
import gradio

def greet(name):
    return "Hello " + name + "!"

demo = gradio.Interface(
    fn = greet,
    inputs = "text",
    outputs = "text"
)

gradio.close_all()

demo.launch(
    server_port = 7860
)
```

Both the code and the embedded Gradio interface will be shown in *Figure 20.1*:

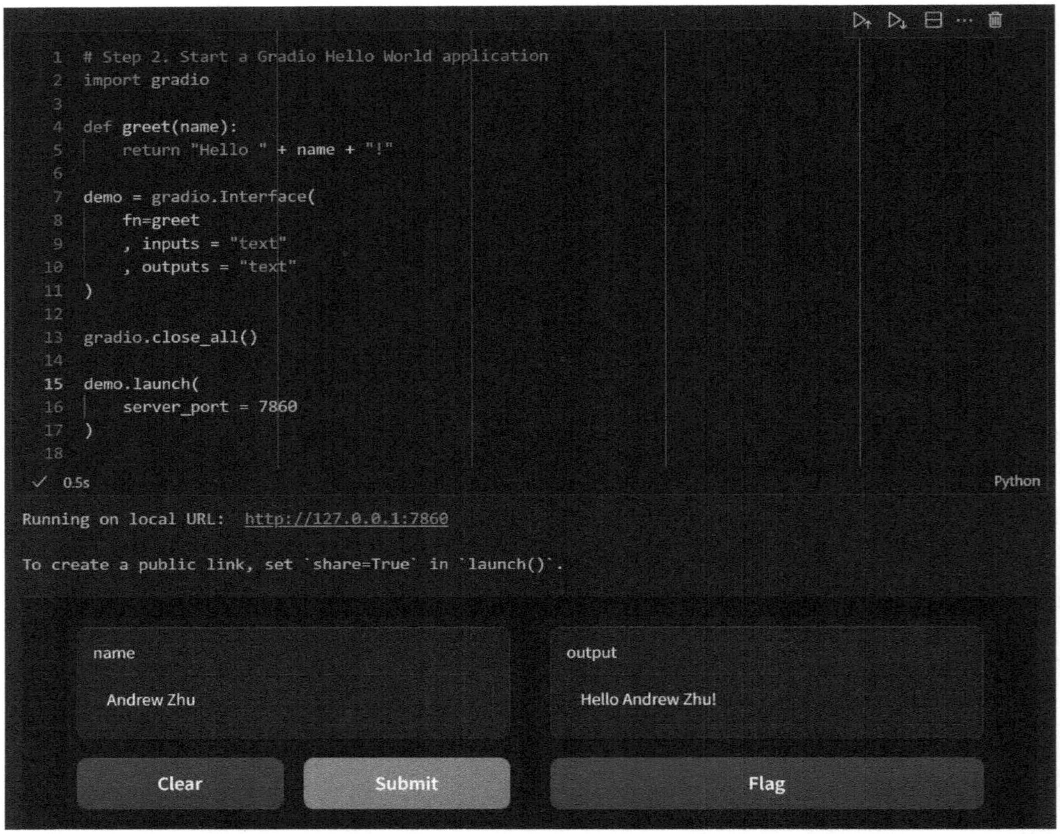

Figure 20.1: Gradio UI embedded in Jupyter Notebook cell

Note that the Jupyter Notebook is running in Visual Studio Code. It also works in Google Colab or independently installed Jupyter Notebook.

Alternatively, we can start a Gradio application from the terminal.

Starting a web application in Jupyter Notebook is good for testing and proof of concept demonstration. When deploying an application, we'd better start it from the terminal.

Create a new file called `gradio_app.py`, and use the same code we used in *step 2*. Use a new port number, such as `7861`, to avoid conflict with the already used `7860`. Then launch the application from the terminal:

```
python gradio_app.py
```

That is all set. Next, let's gain some familiarity with the fundamental building blocks of Gradio.

Gradio fundamentals

The preceding sample code is reformed from the Gradio official quick start tutorial. When we look at the code, lots of details are hidden. We don't know where the Clear button is, we don't specify the Submit button, and we don't know what the Flag button is.

Before using Gradio for any serious applications, we need to understand every line of code and ensure every element is under control.

Instead of using the Interface function to automatically provide the layout, Blocks may provide a better way for us to add interface elements with explicit declaration.

Gradio Blocks

The Interface function provides an abstraction level to easily create quick demos, but there is an abstraction layer. Easy comes with a price. Blocks, on the other hand, is a low-level approach to lay out elements and define data flows. With the help of Blocks, we can precisely control the following:

- The layout of components
- The events that trigger actions
- The direction of the data flow

An example will explain it better:

```python
import gradio
gradio.close_all()

def greet(name):
    return f"hello {name} !"

with gradio.Blocks() as demo:
    name_input = gradio.Textbox(label = "Name")
    output = gradio.Textbox(label = "output box")
    diffusion_btn = gradio.Button("Generate")
    diffusion_btn.click(
        fn = greet,
        inputs = name_input,
        outputs = output
    )

demo.launch(server_port = 7860)
```

The preceding code will produce an interface as shown in *Figure 20.2*:

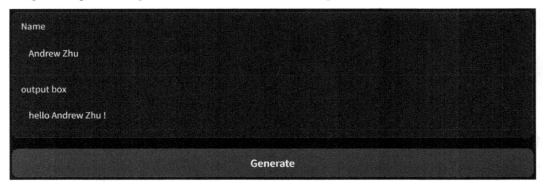

Figure 20.2: Stack Gradio UI using Blocks

All elements under `Blocks` will be shown in the UI. The text for the UI elements is defined by us too. In the `click` event, we defined the `fn` event function, `inputs`, and `outputs`. Finally, launch the application using `demo.launch(server_port = 7860)`.

In line with one of Python's guiding principles: *"Explicit is better than implicit"*, we strive for clarity and simplicity in our code.

Inputs and outputs

The code in the *Gradio Blocks* section uses only one input and one output. We can provide multiple inputs and outputs, as shown in the following code:

```
import gradio
gradio.close_all()

def greet(name, age):
    return f"hello {name} !", f"You age is {age}"

with gradio.Blocks() as demo:
    name_input = gradio.Textbox(label = "Name")
    age_input = gradio.Slider(minimum =0,maximum =100,
        label ="age slider")
    name_output = gradio.Textbox(label = "name output box")
    age_output = gradio.Textbox(label = "age output")
    diffusion_btn = gradio.Button("Generate")
    diffusion_btn.click(
        fn = greet,
        inputs = [name_input, age_input],
        outputs = [name_output, age_output]
```

```
    )

demo.launch(server_port = 7860)
```

The result is shown in *Figure 20.3*:

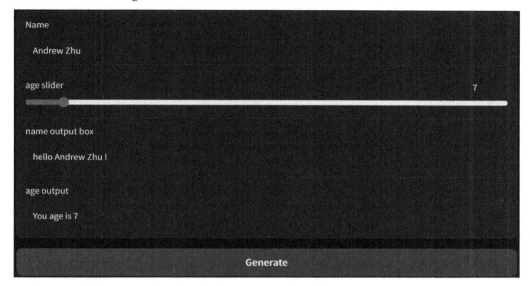

Figure 20.3: Gradio UI with multiple inputs and outputs

Simply stack the element under `with gradio.Blocks() as demo:` and provide inputs and outputs in `list`. Gradio will automatically take the values from the inputs and forward them to the `greet` bind function. The outputs will take the return tuple value from the associated function.

Next, replace the elements with a prompt and output image components. This approach can be applied to build a web-based Stable Diffusion pipeline. However, before proceeding, we need to explore how to integrate a progress bar into our interface.

Building a progress bar

To use a progress bar in Gradio, we can add a `progress` argument to the associated event function. The `Progress` object will be used to track the progress of the function, and it will be displayed to the user as a progress bar.

Here is an example of how to use a progress bar in Gradio.

```
import gradio, time
gradio.close_all()
```

```
def my_function(text, progress=gradio.Progress()):
    for i in range(10):
        time.sleep(1)
        progress(i/10, desc=f"{i}")
    return text

with gradio.Blocks() as demo:
    input = gradio.Textbox()
    output = gradio.Textbox()
    btn = gradio.Button()
    btn.click(
        fn = my_function,
        inputs = input,
        outputs = output
    )

demo.queue().launch(server_port=7860)
```

In the preceding code, we manually update the progress bar with `progress(i/10, desc=f"{i}")`. After each sleep, the progress bar will move forward 10%.

After clicking the **Run** button, the progress bar will appear in the position of the output textbox. We will use a similar approach to apply the progress bar for the Stable Diffusion pipeline in the next section.

Building a Stable Diffusion text-to-image pipeline with Gradio

With all preparations ready, now let's build a Stable Diffusion text-to-image pipeline with Gradio. The UI interface will include the following:

- A prompt input box
- A negative prompt input box
- A button with the `Generate` label
- A progress bar when the `Generate` button is clicked
- An output image

Here is the code that implements these five elements:

```
import gradio
gradio.close_all(verbose = True)
```

```python
import torch
from diffusers import StableDiffusionPipeline

text2img_pipe = StableDiffusionPipeline.from_pretrained(
    "stablediffusionapi/deliberate-v2",
    torch_dtype = torch.float16,
    safety_checker = None
).to("cuda:0")

def text2img(
    prompt:str,
    neg_prompt:str,
    progress_bar = gradio.Progress()
):
    return text2img_pipe(
        prompt = prompt,
        negative_prompt = neg_prompt,
        callback = (
            lambda step,
            timestep,
            latents: progress_bar(step/50,desc="denoising")
        )
    ).images[0]

with gradio.Blocks(
    theme = gradio.themes.Monochrome()
) as sd_app:
    gradio.Markdown("# Stable Diffusion in Gradio")
    prompt = gradio.Textbox(label="Prompt", lines = 4)
    neg_prompt = gradio.Textbox(label="Negative Prompt", lines = 2)
    sd_gen_btn = gradio.Button("Generate Image")
    output_image = gradio.Image()

    sd_gen_btn.click(
        fn = text2img,
        inputs = [prompt, neg_prompt],
        outputs = output_image
    )

sd_app.queue().launch(server_port = 7861)
```

In the preceding code, we first launch the `text2img_pipe` pipeline to VRAM, followed by creating a `text2img` function, which will be called by the Gradio event button. Note the `lambda` expression:

```
callback = (
    lambda step, timestep, latents:
        progress_bar(step/50, desc="denoising")
)
```

We will pass the progress bar into the diffusers denoising loop. Each denoising step will then update the progress bar.

The last part of the code is the Gradio elements `Block` stack. The code also gives Gradio a new theme:

```
...
with gradio.Blocks(
    theme = gradio.themes.Monochrome()
) as sd_app:
...
```

Now you should be able to run the code and generate some images in both Jupyter Notebook and any local web browser.

The progress bar and the result are shown in *Figure 20.4*:

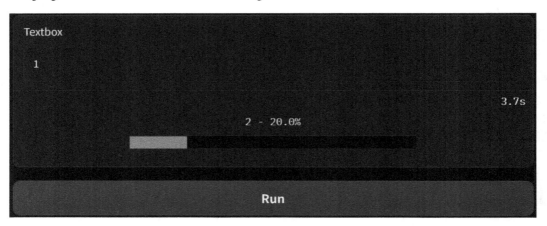

Figure 20.4: Gradio UI with progress bar

You can add more elements and features to this sample application.

Summary

At the time of writing this chapter (December 2023), there isn't much information or sample code to help us get started using diffusers with Gradio. We wrote this chapter to help quickly build up a Stable Diffusion application in Web UI so that we can share the result with others in minutes without touching one line of HTML, CSS, or JavaScript, using pure Python throughout the building process.

This chapter introduced Gradio, what it can do, and why it is popular. We didn't touch on every bit of Gradio; we believe that its official document [1] does this job better. Instead, we used a simple example to explain the backbone of Gradio and what we need to prepare to build a Stable Diffusion Web UI with Gradio.

Finally, we introduced `Blocks`, `inputs`, `outputs`, the progress bar, and event bindings all together and built up a tiny but fully functioning Stable Diffusion pipeline in Gradio.

In the next chapter, we will delve into a relatively complex topic: model fine-tuning and LoRA training.

References

1. Gradio: Build Machine Learning Web Apps — in Python: `https://github.com/gradio-app/gradio`

2. Gradio QuickStart: `https://www.gradio.app/guides/quickstart`

21

Diffusion Model Transfer Learning

This book is mainly focused on using Stable Diffusion with Python, and when doing so, we will need to fine-tune a model for our specific needs. As we discussed in previous chapters, there are many ways to customize the model, such as the following:

- Unlocking UNet to fine-tune all parameters
- Training a textual inversion to add new keyword embeddings
- Locking UNet and training a LoRA model for customized styles
- Training a ControlNet model to guide image generation with control guidance
- Training an adaptor to use the image as one of the guidance embeddings

It is impossible to cover all the model training topics in simply one chapter. Another book would be needed to discuss the details of model training.

Nevertheless, we still want to use this chapter to drill down to the core concepts of model training. Instead of listing sample code on how to fine-tune a diffusion model, or using the scripts from the `Diffusers` package, we want to introduce you to the core concepts of training so that you fully understand the common training process. In this chapter, we will cover the following topics:

- Introducing the foundations of training a model by training a linear model from scratch using PyTorch
- Introducing the Hugging Face Accelerate package to train a model in multiple GPUs
- Building code to train a Stable Diffusion V1.5 LoRA model using PyTorch and Accelerator step by step

By the end of this chapter, you'll be familiar with the overall training process and key concepts, and you'll be able to read sample code from other repositories and build your own training code to customize a model from a pre-trained model.

Writing code to train one model is the best way to learn how to train a model. Let's start work on it.

Technical requirements

Training a model requires more GPU power and VRAM than model inference. Prepare a GPU with at least 8 GB of VRAM – the more, the better. You can also train a model using multiple GPUs.

It is recommended to install the latest version of the following packages:

```
pip install torch torchvision torchaudio
pip install bitsandbytes
pip install transformers
pip install accelerate
pip install diffusers
pip install peft
pip install datasets
```

Here are the specified packages with the versions I used to write the code samples:

```
pip install torch==2.1.2 torchvision==0.16.1 torchaudio==2.1.1
pip install bitsandbytes==0.41.0
pip install transformers==4.36.1
pip install accelerate==0.24.1
pip install diffusers==0.26.0
pip install peft==0.6.2
pip install datasets==2.16.0
```

The training code was tested in the Ubuntu 22.04 x64 version.

Training a neural network model with PyTorch

The target of this section is to build and train one simple neural network model using PyTorch. The model will be a simple one-layer model, with no additional fancy layers. It is simple but with all the elements required to train a Stable Diffusion LoRA, as we will see later in this chapter.

Feel free to skip this section if you are familiar with PyTorch model training. If it is your first time to start training a model, this simple model training will help you thoroughly understand the process of model training.

Before starting, make sure you have installed all the required packages mentioned in the *Technical requirements* section.

Preparing the training data

Let's assume we want to train a model with four weights and output one digital result show, as shown in the following:

$$y = w_1 \times x_1 + w_2 \times x_2 + w_3 \times x_3 + w_4 \times x_4$$

The four weights, w_1, w_2, w_3, w_4, are the model weights we want to have from the training data (Think of these weights as the Stable Diffusion model weight). Because we need to have some real data to train the model, I will use the weights [2,3,4,7] to generate some sample data:

```
import numpy as np
w_list = np.array([2,3,4,7])
```

Let's create 10 groups of input sample data, x_sample; each x_sample is an array with four elements, the same length as the weight:

```
import random
x_list = []
for _ in range(10):
    x_sample = np.array([random.randint(1,100) for _ in range(
        len(w_list))])
    x_list.append(x_sample)
```

In the following section, we will use a neural network model to predict a list of weights; for the sake of training, let's assume that the true weights are unknown after generating the training data.

In the preceding code snippet, we utilize numpy to leverage its dot product operator, @, to compute the output, y. Now, let's generate y_list containing 10 elements:

```
y_list = []
for x_sample in x_list:
    y_temp = x_sample@w_list
    y_list.append(y_temp)
```

You can print x_list and y_list to take a look at the training data.

Our training data is ready; there's no need to download anything else. Next, let's define the model itself and prepare for training.

Preparing for training

Our model could be the world's simplest model ever, a simple linear dot product, as defined in the following code:

```
import torch
import torch.nn as nn
```

```
class MyLinear(nn.Module):
    def __init__(self):
        super().__init__()
        self.w = nn.Parameter(torch.randn(4))

    def forward(self, x:torch.Tensor):
        return self.w @ x
```

The `torch.randn(4)` code is to generate a tensor with a four-weight number. No other code is needed; our NN model is ready now, named `MyLinear`.

To train a model, we will need to initialize it, similar to initializing random weights in an LLM or diffusion model:

```
model = MyLinear()
```

Almost all neural network model training follows these steps:

1. Forward a pass to predict the result.

2. Compute the difference between the predicted result and the ground truth, known as the loss value.

3. Perform backpropagation to calculate the gradient loss value.

4. Update the model parameters.

Before kicking off the training, define a loss function and an optimizer. The loss function, `loss_fn`, will help calculate a loss value based on the predicted result and ground truth result. `optimizer` will be used to update the weights.

```
loss_fn = nn.MSELoss()
optimizer = torch.optim.SGD(model.parameters(), lr = 0.00001)
```

`lr` represents the learning rate, a crucial hyperparameter to set. Determining the best **learning rate** (**lr**) often involves trial and error, depending on the characteristics of your model, dataset, and problem. To find a reasonable learning rate, you need to do the following:

- **Start with a small learning rate**: A common practice is to start with a small learning rate, such as 0.001, and gradually increase or decrease it based on the observed convergence behavior.

- **Use learning rate schedules**: You can use learning rate schedules to adjust the learning rate dynamically during training. One common approach is step decay, where the learning rate decreases after a fixed number of epochs. Another popular method is exponential decay, in which the learning rate decreases exponentially over time. (We won't use it in the world's simplest model.)

Also, don't forget to convert the input and output to the torch Tensor object:

```
x_input = torch.tensor(x_list, dtype=torch.float32)
y_output = torch.tensor(y_list, dtype=torch.float32)
```

All the preparations are done, so let's start training a model.

Training a model

We will set the epoch number to 100, which means looping through our training data 100 times:

```
# start train model
num_epochs = 100
for epoch in range(num_epochs):
    for i, x in enumerate(x_input):
        # forward
        y_pred = model(x)

        # calculate loss
        loss = loss_fn(y_pred,y_output[i])

        # zero out the cached parameter.
        optimizer.zero_grad()

        # backward
        loss.backward()

        # update paramters
        optimizer.step()

    if (epoch+1) % 10 == 0:
        print('Epoch [{}/{}], Loss: {:.4f}'.format(epoch+1,
            num_epochs, loss.item()))

print("train done")
```

Let's break down the preceding code:

- `y_pred = model(x)`: This line applies the model to the current input data sample, `x`, generating a prediction, `y_pred`.

- `loss = loss_fn(y_pred,y_output[i])`: This line calculates the loss (also known as the error or cost) by comparing the predicted output, `y_pred`, with the actual output, `y_output[i]`, using a specified loss function, `loss_fn`.

- `optimizer.zero_grad()`: This line resets the gradients calculated during the backward pass to zero. This is important because it prevents gradient values from carrying over between different samples.

- `loss.backward()`: This line performs the backpropagation algorithm, computing gradients for all parameters with respect to the loss.

- `optimizer.step()`: This line updates the model's parameters based on the computed gradients and the chosen optimization method.

Putting all the code together and running it, we will see the following output:

```
Epoch [10/100], Loss: 201.5572
Epoch [20/100], Loss: 10.8380
Epoch [30/100], Loss: 3.5255
Epoch [40/100], Loss: 1.7397
Epoch [50/100], Loss: 0.9160
Epoch [60/100], Loss: 0.4882
Epoch [70/100], Loss: 0.2607
Epoch [80/100], Loss: 0.1393
Epoch [90/100], Loss: 0.0745
Epoch [100/100], Loss: 0.0398
train done
```

The loss value converges quickly and approaches 0 after 100 epochs. Execute the following code to see the current weight:

```
model.w
```

You can see that the weights update as follows:

```
Parameter containing:
tensor([1.9761, 3.0063, 4.0219, 6.9869], requires_grad=True)
```

This is quite close to [2,3,4,7]! The model was successfully trained to find the right weight numbers.

In the case of Stable Diffusion and multiple GPU training, we can get help from the Hugging Face Accelerate package [4]. Let's start using `Accelerate` next.

Training a model with Hugging Face's Accelerate

Hugging Face's `Accelerate` is a library that provides a high-level API over different PyTorch distributed frameworks, aiming to simplify the process of distributed and mixed-precision training. It is designed to keep changes to your training loop to a minimum and allow the same functions to work for any distributed setup. Let's see what `Accelerate` can bring to the table.

Applying Hugging Face's Accelerate

Let's apply Accelerate to our simple but working model. Accelerate is designed to be used together with PyTorch, so we don't need to change too much code. Here are the steps to use Accelerate to train a model:

1. Generate the default configuration file:

    ```
    from accelerate import utils
    utils.write_basic_config()
    ```

2. Initialize an Accelerate instance, and send the model instance and data to the device managed by Accelerate:

    ```
    from accelerate import Accelerator
    accelerator = Accelerator()
    device = accelerator.device

    x_input.to(device)
    y_output.to(device)
    model.to(device)
    ```

3. Replace loss.backward with accelerator.backward(loss):

    ```
    # loss.backward
    accelerator.backward(loss)
    ```

Next, we will update the training code using Accelerate.

Putting code together

We will keep all the other code the same; here is the complete training code except for the data preparations and model initializing:

```
# start train model using Accelerate
from accelerate import utils
utils.write_basic_config()

from accelerate import Accelerator
accelerator = Accelerator()
device = accelerator.device

x_input.to(device)
y_output.to(device)
model.to(device)
```

```
model, optimizer = accelerator.prepare(
    model, optimizer
)

num_epochs = 100
for epoch in range(num_epochs):
    for i, x in enumerate(x_input):
        # forward
        y_pred = model(x)

        # calculate loss
        loss = loss_fn(y_pred,y_output[i])

        # zero out the cached parameter.
        optimizer.zero_grad()

        # backward
        #loss.backward()
        accelerator.backward(loss)

        # update paramters
        optimizer.step()

    if (epoch+1) % 10 == 0:
        print('Epoch [{}/{}], Loss: {:.4f}'.format(epoch+1,
            num_epochs, loss.item()))

print("train done")
```

Running the preceding code, we should get the same output as when we run the training model without the Hugging Face `Accelerate` library. And the loss value converges as well.

Training a model with multiple GPUs using Accelerate

There are many types of multiple GPU training; in our case, we will use the data parallel style [1]. Simply put, we will load the whole model data into each GPU and split the training data across multiple GPUs.

In PyTorch, we can achieve this with the following code:

```
import torch.nn as nn
from torch.nn.parallel import DistributedDataParallel

model = MyLinear()
ddp_model = DistributedDataParallel(model)
```

```
# Hugging Face Accelerate wraps this operation automatically using the
prepare() function like this:
from accelerate import Accelerator
accelerator = Accelerator()

model = MyLinear()
model = accelerator.prepare(model)
```

For the world's simplest model, we will load the whole model to each GPU and split the 10 groups' training data into 5 groups each. Each GPU will take five groups of data at the same time. After each step, all loss gradient numbers will be merged using the `allreduce` operation. The `allreduce` operation simply added all the loss data from all GPUs, added it up, and then sent it back to each GPU to update the weights, as shown in the following Python code:

```
def allreduce(data):
    for i in range(1, len(data)):
        data[0][:] += data[i].to(data[0].device)
    for i in range(1, len(data)):
        data[i][:] = data[0].to(data[i].device)
```

Accelerate will launch two independent processes to train. To avoid creating two training datasets, let's generate one dataset and save it to local storage using the `pickle` package:

```
import numpy as np
w_list = np.array([2,3,4,7])

import random
x_list = []
for _ in range(10):
    x_sample = np.array([random.randint(1,100)
        for _ in range(len(w_list))]
    )
    x_list.append(x_sample)

y_list = []
for x_sample in x_list:
    y_temp = x_sample@w_list
    y_list.append(y_temp)
train_obj = {
    'w_list':w_list.tolist(),
    'input':x_list,
    'output':y_list
}
```

```
import pickle
with open('train_data.pkl','wb') as f:
    pickle.dump(train_obj,f)
```

Then, wrap the whole model and training code in a `main` function and save it in a new Python file named `train_model_in_2gpus.py`:

```
import torch
import torch.nn as nn
from accelerate import utils
from accelerate import Accelerator

# start a accelerate instance
utils.write_basic_config()
accelerator = Accelerator()
device = accelerator.device

def main():
    # define the model
    class MyLinear(nn.Module):
        def __init__(self):
            super().__init__()
            self.w = nn.Parameter(torch.randn(len(w_list)))

        def forward(self, x:torch.Tensor):
            return self.w @ x

    # load training data
    import pickle
    with open("train_data.pkl",'rb') as f:
        loaded_object = pickle.load(f)
    w_list = loaded_object['w_list']
    x_list = loaded_object['input']
    y_list = loaded_object['output']

    # convert data to torch tensor
    x_input = torch.tensor(x_list, dtype=torch.float32).to(device)
    y_output = torch.tensor(y_list, dtype=torch.float32).to(device)

    # initialize model, loss function, and optimizer
    Model = MyLinear().to(device)
    loss_fn = nn.MSELoss()
    optimizer = torch.optim.SGD(model.parameters(), lr = 0.00001)
```

```
    # wrap model and optimizer using accelerate
    model, optimizer = accelerator.prepare(
        model, optimizer
    )

    num_epochs = 100
    for epoch in range(num_epochs):
        for i, x in enumerate(x_input):
            # forward
            y_pred = model(x)

            # calculate loss
            loss = loss_fn(y_pred,y_output[i])

            # zero out the cached parameter.
            optimizer.zero_grad()

            # backward
            #loss.backward()
            accelerator.backward(loss)

            # update paramters
            optimizer.step()

        if (epoch+1) % 10 == 0:
            print('Epoch [{}/{}], Loss: {:.4f}'.format(epoch+1,
                num_epochs, loss.item()))

    # take a look at the model weights after trainning
    model = accelerator.unwrap_model(model)
    print(model.w)

if __name__ == "__main__":
    main()
```

Then, start the training using this command:

```
accelerate launch --num_processes=2 train_model_in_2gpus.py
```

You should see something like this:

```
Parameter containing:
tensor([1.9875, 3.0020, 4.0159, 6.9961], device='cuda:0', requires_
grad=True)
```

If so, congratulations! You just successfully trained an AI model in two GPUs. With the knowledge you've learned, let's now start to train a Stable Diffusion V1.5 LoRA.

Training a Stable Diffusion V1.5 LoRA

The Hugging Face document provides complete guidance on training a LoRA by calling a pre-defined script [2] provided by Diffusers. However, we don't want to stop at "using" the script. The training code from Diffusers includes a lot of edge-case handling and additional code that is hard to read and learn. In this section, we will write up each line of the training code to fully understand what happens in each step.

In the following sample, we will use eight images with associated captions to train a LoRA. The image and image captions are provided in the `train_data` folder of the code for this chapter.

Our training code structure will be like this:

```
# import packages
import torch
from accelerate import utils
from accelerate import Accelerator
from diffusers import DDPMScheduler,StableDiffusionPipeline
from peft import LoraConfig
from peft.utils import get_peft_model_state_dict
from datasets import load_dataset
from torchvision import transforms
import math
from diffusers.optimization import get_scheduler
from tqdm.auto import tqdm
import torch.nn.functional as F
from diffusers.utils import convert_state_dict_to_diffusers

# train code
def main():
    accelerator = Accelerator(
        gradient_accumulation_steps = gradient_accumulation_steps,
        mixed_precision = "fp16"
    )
    Device = accelerator.device
    ...
    # almost all training code will be land inside of this main
function.

if __name__ == "__main__":
    main()
```

Right below the `main()` function, we initialize the `accelerate` instance. The `Accelerator` instance is initialized with two hyperparameters:

- `gradient_accumulation_steps`: This is the number of training steps to accumulate gradients before we update the model parameters. Gradient accumulation allows you to effectively train with a larger batch size than would be possible with a single GPU, while still fitting the model parameters in memory.

- `mixed_precision`: This specifies the precision to use during training. The `"fp16"` value means that half-precision floating point values will be used for the intermediate computations, which can lead to faster training times and lower memory usage.

The `Accelerator` instance also has an attribute device, which is the device (GPU or CPU) on which the model will be trained. The device attribute can be used to move the model and tensors to the appropriate device before training.

Now, let's start defining hyperparameters.

Defining training hyperparameters

Hyperparameters are parameters that are not learned from the data but, instead, are set before the commencement of the learning process. They are user-defined settings that govern the training process of a machine learning algorithm. In our LoRA training case, we will have the following settings:

```
# hyperparameters
output_dir = "."
pretrained_model_name_or_path  = "runwayml/stable-diffusion-v1-5"
lora_rank = 4
lora_alpha = 4
learning_rate = 1e-4
adam_beta1, adam_beta2 = 0.9, 0.999
adam_weight_decay = 1e-2
adam_epsilon = 1e-08
dataset_name = None
train_data_dir = "./train_data"
top_rows = 4
output_dir = "output_dir"
resolution = 768
center_crop = True
random_flip = True
train_batch_size = 4
gradient_accumulation_steps = 1
num_train_epochs = 200
# The scheduler type to use. Choose between ["linear", "cosine", #
"cosine_with_restarts", "polynomial","constant", "constant_with_
```

```
# warmup"]
lr_scheduler_name = "constant" #"cosine"#
max_grad_norm = 1.0
diffusion_scheduler = DDPMScheduler
```

Let's break down the preceding settings:

- `output_dir`: This is the directory where the model outputs will be saved.

- `pretrained_model_name_or_path`: This is the name or path of the pretrained model to be used as the starting point for training.

- `lora_rank`: This is the number of layers in the **low-rank adaptation module (LoRa)** used to fine-tune the pretrained model. A higher rank allows for more complex adjustments, but it also requires more training data and can be computationally expensive. Generally, ranks below 32 might not be effective enough, while ranks above 256 might be overkill for most tasks. In our case, since we use only eight images to train the LoRA, setting the rank to 4 is enough.

- `lora_alpha`: This, conversely, controls the strength of the updates made to the pretrained model's weights during fine-tuning. Specifically, the weight changes generated during fine-tuning are multiplied by a scaling factor equal to Alpha divided by Rank, before being added back to the original model weights. Therefore, increasing Alpha relative to Rank. Setting Alpha equal to Rank is a common starting practice.

- `learning_rate`: This parameter controls how quickly the model learns from its mistakes during training. Specifically, it sets the step size for each iteration, determining how aggressively the model adjusts its parameters to minimize the `loss` function.

- `adam_beta1` and `adam_beta2`: These are the parameters used in the Adam optimizer to control the decay rates of the moving averages of the gradient and squared gradient, respectively.

- `adam_weight_decay`: This is the weight decay used in the Adam optimizer to prevent overfitting.

- `adam_epsilon`: This is a small value added to the denominator for numerical stability in the Adam optimizer.

- `dataset_name`: This is the name of the dataset to be used for training. Particularly, this is the Hugging Face dataset ID, such as `lambdalabs/pokemon-blip-captions`.

- `train_data_dir`: This is the directory where the training data is stored.

- `top_rows`: This is the number of rows used for training. It is used to select the top rows for training; if you have a dataset with 1,000 rows, set it to 8 to train the training code with the top 8 rows.

- `output_dir`: This is the directory where the outputs will be saved during training.

- `resolution`: This is the resolution of the input images.

- `center_crop`: This is a Boolean flag indicating whether to perform center cropping on the input images.

- `random_flip`: This is a Boolean flag indicating whether to perform random horizontal flipping on the input images.

- `train_batch_size`: This is the batch size used during training.

- `gradient_accumulation_steps`: This is the number of training steps to accumulate gradients before updating the model parameters.

- `num_train_epochs`: This is the number of training epochs to perform.

- `lr_scheduler_name`: This is the name of the learning rate scheduler to use.

- `max_grad_norm`: This is the maximum norm of the gradients to clip to prevent exploding gradients.

- `diffusion_scheduler`: This is the name of the diffusion scheduler to use.

Preparing the Stable Diffusion components

When training a LoRA, the process involves inference, adding the loss value, and backpropagation - a procedure reminiscent of the inference process. To facilitate this, let's use the `StableDiffusion-Pipeline` from `Diffusers` package to get `tokenizer`, `text_encoder`, `vae`, and `unet`:

```
noise_scheduler = DDPMScheduler.from_pretrained(
    pretrained_model_name_or_path, subfolder="scheduler")
weight_dtype = torch.float16
pipe = StableDiffusionPipeline.from_pretrained(
    pretrained_model_name_or_path,
    torch_dtype = weight_dtype
).to(device)
tokenizer, text_encoder = pipe.tokenizer, pipe.text_encoder
vae, unet = pipe.vae, pipe.unet
```

During LoRA training, those components will facilitate the forward pass, but their weights won't be updated during backpropagation, so we need to set `requires_grad_` to `False`, as shown here:

```
# freeze parameters of models, we just want to train a LoRA only
unet.requires_grad_(False)
vae.requires_grad_(False)
text_encoder.requires_grad_(False)
```

The LoRA weights are the part we want to train; let's use PEFT's [3] `LoraConfig` to initialize the LoRA configurations.

PEFT is a library developed by Hugging Face that provides parameter-efficient ways to adapt large pre-trained models to specific downstream applications. The key idea behind PEFT is to fine-tune only a small fraction of a model's parameters instead of fine-tuning all of them, resulting in significant savings in terms of computation and memory usage. This makes it possible to fine-tune very large models even on consumer hardware with limited resources.

LoRA is one of the PEFT methods supported by the PEFT library. With LoRA, instead of updating all the weights of a given layer during fine-tuning, only a low-rank approximation of the weight updates is learned, reducing the number of additional parameters required per layer. This approach allows you to fine-tune just 0.16% of the total parameters of a model while achieving similar performance to full fine-tuning.

To use LoRA with a pre-trained transformer model, you need to instantiate a `LoraConfig` object and pass it to the appropriate component of your model. The `LoraConfig` class has several attributes that control its behavior, including the dimension/rank of the decomposition, dropout rates, and other hyperparameters. Once configured, you can then train your model using standard techniques, such as gradient descent. Here is the code to create a LoRA configuration object:

```
# configure LoRA parameters use PEFT
unet_lora_config = LoraConfig(
    r = lora_rank,
    lora_alpha = lora_alpha,
    init_lora_weights = "gaussian",
    target_modules = ["to_k", "to_q", "to_v", "to_out.0"]
)
```

Next, let's add the LoRA adapter to the UNet model using the `unet_lora_config` configuration:

```
# Add adapter and make sure the trainable params are in float32.
unet.add_adapter(unet_lora_config)
for param in unet.parameters():
    # only upcast trainable parameters (LoRA) into fp32
    if param.requires_grad:
        param.data = param.to(torch.float32)
```

Inside the `for` loop, if the parameters require gradients (i.e., they are trainable), their data type is explicitly cast to `torch.float32`. This ensures that only the trainable parameters are in the `float32` format for efficient training.

Loading the training data

Let's load up some data using the following code:

```
if dataset_name:
    # Downloading and loading a dataset from the hub. data will be
    # saved to ~/.cache/huggingface/datasets by default
```

```
    dataset = load_dataset(dataset_name)
else:
    dataset = load_dataset(
        "imagefolder",
        data_dir = train_data_dir
    )

train_data = dataset["train"]
dataset["train"] = train_data.select(range(top_rows))

# Preprocessing the datasets. We need to tokenize inputs and targets.
dataset_columns = list(dataset["train"].features.keys())
image_column, caption_column = dataset_columns[0],dataset_columns[1]
```

Let's break down the preceding code:

- `if dataset_name:`: If `dataset_name` is provided, the code tries to load a dataset from Hugging Face's dataset hub using the `load_dataset` function. If no `dataset_name` is provided, it assumes that the dataset is stored locally and loads it using the `imagefolder` dataset type.

- `train_data = dataset["train"]`: The train split of the dataset is assigned to the `train_data` variable.

- `dataset["train"] = train_data.select(range(top_rows))`: The first top rows of the train dataset are selected and assigned back to the train split of the dataset. This is useful when working with a small subset of the dataset for faster experimentation.

- `dataset_columns = list(dataset["train"].features.keys())`: The keys of the `dataset["train"]` feature dictionary are extracted and assigned to the `dataset_columns` variable. These keys represent the image and caption columns in the dataset.

- `image_column, caption_column = dataset_columns[0], dataset_columns[1]`: The first and second columns are assigned to the `image_column` and `caption_column` variables, respectively. This assumes that the dataset has exactly two columns – the first for images and the second for captions.

We will need a function to convert the input text to token IDs; we define the function like this:

```
def tokenize_captions(examples, is_train=True):
    '''Preprocessing the datasets.We need to tokenize input captions
and transform the images.'''
    captions = []
    for caption in examples[caption_column]:
        if isinstance(caption, str):
            captions.append(caption)
```

```
inputs = tokenizer(
    captions,
    max_length = tokenizer.model_max_length,
    padding = "max_length",
    truncation = True,
    return_tensors = "pt"
)
return inputs.input_ids
```

And then, we train the data transform pipeline:

```
# Preprocessing the datasets.
train_transforms = transforms.Compose(
    [
        transforms.Resize(
            resolution,
            interpolation=transforms.InterpolationMode.BILINEAR
        ),
        transforms.CenterCrop(resolution) if center_crop else
            transforms.RandomCrop(resolution),
        transforms.RandomHorizontalFlip() if random_flip else
            transforms.Lambda(lambda x: x),
        transforms.ToTensor(),
        transforms.Normalize([0.5], [0.5]) # [0,1] -> [-1,1]
    ]
)
```

The preceding code defines a set of image transformations that will be applied to the training dataset during the training of a machine learning or deep learning model. These transformations are defined using the transforms module from the PyTorch library.

Here's a breakdown of what each line does:

- transforms.Compose(): This is a function that "chains" multiple transformations together. It takes a list of transformation functions as input and applies them in order.

- transforms.Resize(resolution, interpolation=transforms. InterpolationMode.BILINEAR): This line resizes the image to the given resolution pixels while keeping the aspect ratio. The interpolation method used is bilinear interpolation.

- transforms.CenterCrop(resolution) if center_crop else transforms. RandomCrop(resolution): This line crops the image to a square of resolution x resolution. If center_crop is True, the crop is taken from the center of the image. If center_crop is False, the crop is taken randomly.

- `transforms.RandomHorizontalFlip() if random_flip else transforms.Lambda(lambda x: x)`: This line horizontally flips the image randomly with a probability of 0.5. If `random_flip` is `False`, it leaves the image unchanged.

- `transforms.ToTensor()`: This line converts the image from a PIL image or NumPy array to a PyTorch tensor.

- `transforms.Normalize([0.5], [0.5])`: This line scales the pixel values of the image between -1 and 1. It is commonly used to normalize image data before passing it to a neural network.

By chaining these transformations together using `transforms.Compose`, you can easily preprocess your image data and apply multiple transformations to your dataset.

We need the following code to use the chained transformation object:

```
def preprocess_train(examples):
    '''prepare the train data'''
    images = [image.convert("RGB") for image in examples[
        image_column]]
    examples["pixel_values"] = [train_transforms(image)
        for image in images]
    examples["input_ids"] = tokenize_captions(examples)
    return examples


# only do this in the main process
with accelerator.main_process_first():
    # Set the training transforms
    train_dataset = dataset["train"].with_transform(preprocess_train)

def collate_fn(examples):
    pixel_values = torch.stack([example["pixel_values"]
        for example in examples])
    pixel_values = pixel_values.to(memory_format = \
        torch.contiguous_format).float()
    input_ids = torch.stack([example["input_ids"]
        for example in examples])
    return {"pixel_values": pixel_values, "input_ids": input_ids}


# DataLoaders creation:
train_dataloader = torch.utils.data.DataLoader(
    train_dataset,
    shuffle = True
    collate_fn = collate_fn
    batch_size = train_batch_size
)
```

The preceding code first defines a function called `preprocess_train`, which preprocesses the train data. It first converts the images to the RGB format, and then it applies a series of image transformations (resize, center/random crop, random horizontal flip, and normalization) to them using the `train_transforms` object. It then tokenizes the input captions using the `tokenize_captions` function. The resulting preprocessed data is added to the `examples` dictionary as the `pixel_values` and `input_ids` keys.

The with `accelerator.main_process_first()` line is used to ensure that the code inside the block is executed only in the main process. In this case, it sets the training transforms for `train_dataset`.

The `collate_fn` function is used to collate the dataset examples into a batch to be fed to the model. It takes a list of examples and stacks `pixel_values` and `input_ids` together. The resulting tensors are then converted to the `float32` format and returned as a dictionary.

Finally, `train_dataloader` is created using the `torch.utils.data.DataLoader` class, which loads `train_dataset` with the specified batch size, shuffle, and collate functions.

In PyTorch, DataLoader is a utility class that abstracts the process of loading data in batches for training or evaluation. It is used to load data in batches, which are sequences of data points used to train a machine learning model.

In the provided code, `train_dataloader` is an instance of PyTorch's `DataLoader` class. It is used to load the training data in batches. More specifically, it loads the data from `train_dataset` in batches of a predefined batch size, shuffles the data for each epoch, and applies a user-defined `collate_fn` function to preprocess the data before feeding it to the model.

`train_dataloader` is necessary for the efficient training of the model. By loading data in batches, it allows the model to process multiple data points in parallel, which can significantly reduce training time. Additionally, shuffling the data for each epoch helps prevent overfitting by ensuring that the model sees different data points in each epoch.

In the provided code, the `collate_fn` function is used to preprocess the data before it is fed to the model. It takes a list of examples and returns a dictionary containing the pixel values and input IDs for each example. The `collate_fn` function is applied to each batch of data by `DataLoader` before it is fed to the model. This allows for more efficient processing of the data by applying the same preprocessing steps to each batch of data.

Defining the training components

To prepare and define the training components, let's first initialize an AdamW optimizer. AdamW is an optimization algorithm to train machine learning models. It is a variant of the popular Adam optimizer, which uses adaptive learning rates for each model parameter. The AdamW optimizer is similar to the Adam optimizer, but it includes an additional weight decay term in the gradient update step. This weight decay term is added to the gradient of the loss function during optimization, which helps to prevent overfitting by adding a regularization term to the loss function.

We can initialize an AdamW optimizer using the following code:

```
# initialize optimizer
lora_layers = filter(lambda p: p.requires_grad, unet.parameters())
optimizer = torch.optim.AdamW(
    lora_layers,
    lr = learning_rate,
    betas = (adam_beta1, adam_beta2),
    weight_decay = adam_weight_decay,
    eps = adam_epsilon
)
```

The filter function is used to iterate through all the parameters of the unet model and selects only those parameters that require gradient computation. The filter function returns a generator object that contains the parameters that require gradient computation. This generator object is assigned to the lora_layers variable, which will be used to optimize the model parameters during training.

The AdamW optimizer is initialized with the following hyperparameters:

- lr: The learning rate, which controls the step size at each iteration while moving toward a minimum of a loss function

- betas: A tuple containing the exponential decay rates for the moving average of the gradient (β1) and the squared gradient (β2)

- weight_decay: The weight decay term added to the gradient of the loss function during optimization

- eps: A small value added to the denominator to improve numerical stability

Second, we define a learning rate scheduler – lr_scheduler. Instead of defining one manually, we can use the get_scheduler function provided by the Diffusers package (from diffusers. optimization import get_scheduler):

```
# learn rate scheduler from diffusers's get_scheduler
lr_scheduler = get_scheduler(
    lr_scheduler_name,
    optimizer = optimizer
)
```

This code creates a learning rate scheduler object using the get_scheduler function from the Diffusers library. The learning rate scheduler determines how the learning rate (i.e., the step size in gradient descent) changes during training.

The `get_scheduler` function takes two arguments:

- `lr_scheduler_name`: The name of the learning rate scheduler algorithm to use. In our sample, the name is `constant`, defined at the beginning of the code.

- `optimizer`: The PyTorch optimizer object that the learning rate scheduler will be applied to. This is the `AdamW` optimizer we just initialized.

We have just prepared all the elements to kick off training and we have written lots of code to prepare the dataset, although the actual training code isn't that long. Let's write the training code next.

Training a Stable Diffusion V1.5 LoRA

Training a LoRA will usually take a while, and we'd better create a progress bar to track the training progress:

```python
# set step count and progress bar
max_train_steps = num_train_epochs*len(train_dataloader)
progress_bar = tqdm(
    range(0, max_train_steps),
    initial = 0,
    desc = "Steps",
    # Only show the progress bar once on each machine.
    Disable = not accelerator.is_local_main_process,
)
```

Here is the core training code:

```python
# start train
for epoch in range(num_train_epochs):
    unet.train()
    train_loss = 0.0
    for step, batch in enumerate(train_dataloader):
        # step 1. Convert images to latent space
        # latents = vae.encode(batch["pixel_values"].to(
            dtype=weight_dtype)).latent_dist.sample()
        latents = latents * vae.config.scaling_factor

        # step 2. Sample noise that we'll add to the latents,
        latents provide the shape info.
        noise = torch.randn_like(latents)

        # step 3. Sample a random timestep for each image
        batch_size = latents.shape[0]
        timesteps = torch.randint(
            low = 0,
```

```
        high = noise_scheduler.config.num_train_timesteps,
        size = (batch_size,),
        device = latents.device
)
timesteps = timesteps.long()


# step 4. Get the text embedding for conditioning
encoder_hidden_states = text_encoder(batch["input_ids"])[0]


# step 5. Add noise to the latents according to the noise
# magnitude at each timestep
# (this is the forward diffusion process),
# provide to unet to get the prediction result
noisy_latents = noise_scheduler.add_noise(
    latents, noise, timesteps)


# step 6. Get the target for loss depend on the prediction
# type
if noise_scheduler.config.prediction_type == "epsilon":
    target = noise
elif noise_scheduler.config.prediction_type == "v_prediction":
    target = noise_scheduler.get_velocity(
        latents, noise, timesteps)
else:
    raise ValueError(f"Unknown prediction type {
        noise_scheduler.config.prediction_type}")


# step 7. Predict the noise residual and compute loss
model_pred = unet(noisy_latents, timesteps,
    encoder_hidden_states).sample


# step 8. Calculate loss
loss = F.mse_loss(model_pred.float(), target.float(),
    reduction="mean")


# step 9. Gather the losses across all processes for logging
# (if we use distributed training).
avg_loss = accelerator.gather(loss.repeat(
    train_batch_size)).mean()
train_loss += avg_loss.item() / gradient_accumulation_steps


# step 10. Backpropagate
accelerator.backward(loss)
if accelerator.sync_gradients:
```

```
            params_to_clip = lora_layers
            accelerator.clip_grad_norm_(params_to_clip, max_grad_norm)

    optimizer.step()
    lr_scheduler.step()
    optimizer.zero_grad()

    # step 11. check optimization step and update progress bar
    if accelerator.sync_gradients:
        progress_bar.update(1)
        train_loss = 0.0

    logs = {"epoch": epoch, "step_loss": loss.detach().item(),
        "lr": lr_scheduler.get_last_lr()[0]}
    progress_bar.set_postfix(**logs)
```

The preceding code is a typical training loop for Stable Diffusion model training. Here's a breakdown of what each part of the code does:

- The outer loop (`for epoch in range(num_train_epochs)`) iterates over the number of training epochs. An epoch is one complete pass through the entire training dataset.

- `unet.train()` sets the model to training mode. This is important because some layers, such as dropout and batch normalization, behave differently during training and testing. In the training phase, these layers behave differently than in the evaluation phase. For example, dropout layers will drop out nodes with a certain probability during training to prevent overfitting, but they will not drop out any nodes during evaluation. Similarly, `BatchNorm` layers will use batch statistics during training, but will use accumulated statistics during evaluation. So, if you don't call `unet.train()`, these layers will not behave correctly for the training phase, which could lead to incorrect results.

- The inner loop (`for step, batch in enumerate(train_dataloader)`) iterates over the training data. `train_dataloader` is a `DataLoader` object that provides batches of training data.

- In *step 1*, the model encodes the input images into a latent space using a **Variational Autoencoder (VAE)**. The latent distribution is then sampled to get the latent vectors (`latents`), which are scaled by a factor.

- In *step 2*, random noise is added to the latent vectors. This noise is sampled from a standard normal distribution and has the same shape as the latent vectors.

- In *step 3*, random timesteps are sampled for each image in the batch. This is part of a time-dependent noise addition process.

- In *step 4*, the text encoder is used to get the text embedding for conditioning.

- In *step 5*, noise is added to the latent vectors according to the noise magnitude at each timestep.

- In *step 6*, the target for the loss calculation is determined based on the prediction type. It can be either the noise or the velocity of the noise.

- In *steps 7* and *8*, The model makes a prediction using the noisy latent vectors, the timesteps, and the text embeddings. The loss is then calculated as the mean squared error between the model's prediction and the target.

- In *step 9*, the loss is gathered across all processes for logging. This is necessary in the case of distributed training, where the model is trained on multiple GPUs. So that we can see the loss value changes in the middle of a training process.

- In *step 10*, the gradients of the loss with respect to the model parameters are computed (`accelerator.backward(loss)`), and the gradients are clipped if necessary. This is to prevent the gradients from becoming too large, which can cause numerical instability. The optimizer updates the model parameters based on the gradients (`optimizer.step()`), and the learning rate scheduler updates the learning rate (`lr_scheduler.step()`). The gradients are then reset to zero (`optimizer.zero_grad()`).

- In *step 11*, if the gradients are synchronized, the training loss is reset to zero and the progress bar is updated.

- Finally, the training loss, learning rate, and current epoch are logged to monitor the training process. The progress bar is updated with these logs.

Once you understand the preceding steps, you can not only train a Stable Diffusion LoRA but also train any other models.

Lastly, we will need to save the LoRA we just trained:

```
# Save the lora layers
accelerator.wait_for_everyone()
if accelerator.is_main_process:
    unet = unet.to(torch.float32)

    unwrapped_unet = accelerator.unwrap_model(unet)
    unet_lora_state_dict = convert_state_dict_to_diffusers(
        get_peft_model_state_dict(unwrapped_unet))

    weight_name = f"""lora_{pretrained_model_name_or_path.split('/')
[-1]}_rank{lora_rank}_s{max_train_steps}_r{resolution}_{diffusion_
scheduler.__name__}_{formatted_date}.safetensors"""
    StableDiffusionPipeline.save_lora_weights(
        save_directory = output_dir,
        unet_lora_layers = unet_lora_state_dict,
        safe_serialization = True,
```

```
        weight_name = weight_name
    )

accelerator.end_training()
```

Let's break down the preceding code:

- `accelerator.wait_for_everyone()`: This line is used in distributed training to make sure all processes have reached this point in the code. It's a synchronization point.

- `if accelerator.is_main_process:`: This checks whether the current process is the main one. In distributed training, you typically only want to save the model once, not once for each process.

- `unet = unet.to(torch.float32)`: This line converts the data type of the model's weights to `float32`. This is typically done to save memory, as `float32` uses less memory than `float64` but still provides sufficient precision for most deep learning tasks.

- `unwrapped_unet = accelerator.unwrap_model(unet)`: This unwraps the model from the accelerator, which is a wrapper used for distributed training.

- `unet_lora_state_dict = convert_state_dict_to_diffusers(get_peft_model_state_dict(unwrapped_unet))`: This line gets the state dictionary of the model, which contains the weights of the model, and then converts it to a format suitable for Diffusers.

- `weight_name = f"lora_{pretrained_model_name_or_path.split('/')[-1]}_rank{lora_rank}_s{max_train_steps}_r{resolution}_{diffusion_scheduler.__name__}_{formatted_date}.safetensors"`: This line creates a name for the file where the weights will be saved. The name includes various details about the training process.

- `StableDiffusionPipeline.save_lora_weights(...)`: This line saves the weights of the model to a file. The `save_directory` argument specifies the directory where the file will be saved, `unet_lora_layers` is the state dictionary of the model, `safe_serialization` indicates that the weights should be saved in a way that is safe to load later, and `weight_name` is the name of the file.

- `accelerator.end_training()`: This line signals the end of the training process. This is typically used to clean up resources used during training.

We have the complete training code in the associated code folder for this chapter, named `train_sd16_lora.py`. We are not done yet; we still need to kick off the training using the `accelerator` command instead of entering `python py_file.py` directly.

Kicking off the training

If you have one GPU, simply run the following command:

```
accelerate launch --num_processes=1 ./train_sd16_lora.py
```

For two GPUs, increase --num_processes to 2, like this:

```
accelerate launch --num_processes=2 ./train_sd16_lora.py
```

If you have more than two GPUs and want to train on assigned GPUs (e.g., you have three GPUs and want the training code run on the second and third GPUs), use this command:

```
CUDA_VISIBLE_DEVICES=1,2 accelerate launch --num_processes=2 ./train_
sd16_lora.py
```

To use the first and third GPUs, simply update the CUDA_VISIBLE_DEVICES settings to 0,2, like this:

```
CUDA_VISIBLE_DEVICES=0,2 accelerate launch --num_processes=2 ./train_
sd16_lora.py
```

Verifying the result

This is the most exciting moment to witness the power of model training. First, let's load up the LoRA but set its weight to 0.0 with adapter_weights = [0.0]:

```
from diffusers import StableDiffusionPipeline
import torch
from diffusers.utils import make_image_grid
from diffusers import EulerDiscreteScheduler

lora_name = "lora_file_name.safetensors"
lora_model_path = f"./output_dir/{lora_name}"

device = "cuda:0"
pipe = StableDiffusionPipeline.from_pretrained(
    "runwayml/stable-diffusion-v1-5",
    torch_dtype = torch.bfloat16
).to(device)

pipe.load_lora_weights(
    pretrained_model_name_or_path_or_dict=lora_model_path,
    adapter_name = "az_lora"
)
```

```
prompt = "a toy bike. macro photo. 3d game asset"
nagtive_prompt = "low quality, blur, watermark, words, name"

pipe.set_adapters(
    ["az_lora"],
    adapter_weights = [0.0]
)

pipe.scheduler = EulerDiscreteScheduler.from_config(
    pipe.scheduler.config)

images = pipe(
    prompt = prompt,
    nagtive_prompt = nagtive_prompt,
    num_images_per_prompt = 4,
    generator = torch.Generator(device).manual_seed(12),
    width = 768,
    height = 768,
    guidance_scale = 8.5
).images

pipe.to("cpu")
torch.cuda.empty_cache()
make_image_grid(images, cols = 2, rows = 2)
```

Running the preceding code, we will get the images shown in *Figure 21.1*:

Figure 21.1: A toy bike, a macro photo, a 3D game asset, and an image generated without using LoRA

The result is not that good. Now, let's enable the trained LoRA with `adapter_weights = [1.0]`. Run the code again, and you should see the images shown in *Figure 21.2*:

Figure 21.2: A toy bike, a macro photo, a 3D game asset, and an image generated with LoRA training using eight images

The result is way better than the images without using the LoRA! If you see similar results, congratulations!

Summary

This has been a long chapter, but learning about the power of model training is worth the length. Once we have mastered the training skill, we can train any models based on our needs. The whole training process isn't easy, as there are so many details and trivial things to deal with. However, writing the training code is the only way to fully understand how model training works; considering the fruitful outcome, it is worth spending time to figure it out from the bottom up.

Due to the length limitation of one chapter, I can only cover the entire LoRA training process, but once you succeed with LoRA training, you can find more training samples from Diffusers, change the code based on your specific needs, or simply write your training code, especially if you are working on a new model's architecture.

In this chapter, we began by training one simple model; the model itself isn't that interesting, but it helped you to understand the core steps of model training using PyTorch. Then, we moved on to leverage the Accelerator package to train a model in multiple GPUs. Finally, we touched on the real Stable Diffusion model and trained a full-functioning LoRA, using simply eight images.

In the next and final chapter, we'll discuss something less technical, AI, and its relationship with us, privacy, and how to keep pace with its fast-changing advancements.

References

1. What is **Distributed Data Parallel** (DDP): `https://pytorch.org/tutorials/beginner/ddp_series_theory.html`

2. Launch the LoRA training script: `https://huggingface.co/docs/diffusers/en/training/lora#launch-the-script`

3. Hugging Face PEFT: `https://huggingface.co/docs/peft/en/index`

4. Hugging Face Accelerate: `https://huggingface.co/docs/accelerate/en/index`

22

Exploring Beyond Stable Diffusion

The realm of Stable Diffusion is in a constant state of flux, with innovative models, methodologies, and research papers surfacing daily. Throughout the course of writing this book, the Stable Diffusion community has seen remarkable growth. Given the dynamic nature of this field, inevitably some developments are not covered within these pages.

In the process of writing this book and delving into the intricacies of Stable Diffusion, I've often been asked, "*How do you begin to understand this complex subject?*" In this concluding chapter, I aim to share my learning journey and provide insights to help you stay abreast of the latest developments in Stable Diffusion and AI.

In this chapter, we will discuss the following:

- **What sets this AI wave apart**: Understanding the unique characteristics of the current AI revolution

- **The enduring value of mathematics and programming**: Emphasizing the importance of core skills in the rapidly changing AI landscape

- **Staying current with AI innovations**: Tips and strategies to keep up with the latest AI breakthroughs

- **Cultivating responsible, ethical, private, and secure AI**: Exploring the best practices to develop AI that aligns with societal values and safety standards

- **Our evolving relationship with AI**: Reflecting on the implications of AI for individuals, organizations, and society as a whole

I hope that this chapter serves as a valuable resource for those eager to expand their knowledge of Stable Diffusion and AI. Curiosity is the key to unlocking deeper understanding and exploration in this exciting field.

What sets this AI wave apart

In March 2016, AlphaGo [1] made history when it defeated the world-famous Go player Lee Sedol in a five-game match. This was a significant event because Go is a game that requires strategic thinking and intuition and it has been considered impossible for computers to master due to its complexity. AlphaGo's victory was a testament to the advancements in AI and machine learning.

AlphaGo's success was based on a combination of deep neural networks and Monte Carlo tree search techniques. It was trained on thousands of professional Go games to learn patterns and strategies. Then, it played many games against itself to improve its skills and understanding of the game.

This achievement marked a major milestone in the development of AI, demonstrating that machines can now outperform humans in tasks that require deep understanding and strategic decision-making.

I was watching these games live and was astonished by the power of the machine. However, the model that powers AlphaGo is not short of limitations. Here are a few:

- **Specificity to Go**: AlphaGo is specifically designed to play the game of Go. It doesn't have the ability to transfer its knowledge to other games or domains. If we added one more row to the game board, AlphaGo would not be able to function as it should.

- **Explainability**: It's difficult to understand why AlphaGo makes certain decisions, which can make it hard to trust or rely on its output in critical situations.

AlphaGo was trained only with GoPlay data, so its data scope is quite limited, which is the root cause of its "specification" feature. It is similar to **Convolution Neural Network (CNN)**-based image classification models. These models are trained by a set of predefined data; hence, they can only be performed on the scoped input data.

In 2017, the paper *"Attention Is All You Need"* [2] introduced the transformer model. The authors demonstrated the effectiveness of the transformer model in a specific unsupervised task called machine translation. They trained the model to translate sentences from one language to another without any aligned sentence pairs or explicit supervision. Instead, they used an encoder-decoder structure to predict the next word with probability. In other words, the next "word" or "token" is the training label for the input, so the model tries to learn patterns or structures present in data without any explicit guidance about what to learn.

The transformer model itself is no doubt still important (at least at the time of writing). But the idea and the implementation of training a model without predefined labels is genius.

In recent years, some models have used only a decoder to train a model. Notably, GPT-3 uses a decoder-only architecture to generate text. Some other visual models use an attention mechanism to replace the CNN structure – for example, a **Vision Transformer (ViT)** [3] and the **Swin transformer** [4].

In the case of Stable Diffusion, the model presented in this book integrates the attention mechanism within its UNet architecture as discussed in *Chapter 4* and *Chapter 5*. Stable Diffusion can take any image and captioned pair for training, without a limitation in its data scope. If we have enough hardware power, we can provide all images of the world and their associated description text to train a super diffusion model.

As the Sora model from OpenAI has shown, with enough video data, associated description, massive GPU power, and a diffusion transformer-based model, a model can generate a video to somewhat mimic the real world.

At the time of writing, we don't know what the limit is of this attention-based auto-learning architecture.

The enduring value of mathematics and programming

As we witness the power demonstrated by AI, some might argue that, in the future, there will be no need to learn programming or math, as we can delegate any tasks to AI. However, this is far from the truth. This wave of AI revolution opens a new door to the future, but fundamentally, AI is still a program running on silicon chips, and it requires human beings to provide wisdom and knowledge.

Current AI technology has been developed based on mathematical models such as probability theory, statistics, and linear algebra, which are essential for AI algorithms. For example, the latent-based diffusion model (Stable Diffusion) is an algorithm based on neural networks, which are inspired by the structure of the human brain. The most important part of deep learning is backpropagation, which is essentially calculus. Therefore, AI cannot exist without mathematics.

In regards to programming skills, GPT and Diffusion models do not make programming skills redundant; on the contrary – programming finds new areas to conquer. Those who contribute to AI development are also those who write most of the code.

Let's assume you are misled by certain self-proclaimed experts and give up pursuing your programming capability. Several years later, when you open any GitHub AI projects, you will not only be unable to make any contributions but also won't be able to read the code at all.

Mathematic knowledge and programming skills will never become obsolete. They may change their forms over time, but the core concepts remain unchanged.

As you are reading a book about using Stable Diffusion with Python, I bet you won't be satisfied with just being able to use the model and need to know how it works internally. To understand it, the best way is to create it, as Richard Feynman [7] said on his blackboard [5], as shown in *Figure 22.1*:

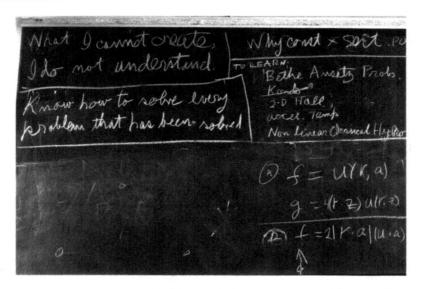

Figure 22.1 – Richard Feynman's blackboard – "What I cannot create, I do not understand"

It is also down to personal experience; I can only fully understand a topic after implementing it. As the old saying goes, we can't learn to swim by reading a book about swimming.

However, implementing a model will require an understanding of the theory, which requires an understanding of related mathematics and the programming skills to convert a complex formula to executable bits.

Let's assume we have set our mind on learning more about AI – where do we start?

Staying current with AI innovations

As the transformer model has largely transformed the landscape, we can't find the newest learning material from the Amazon bookstore, especially books written before 2022. It is important to find the most relevant high-quality information. Here are some channels that might be useful:

- **Follow the renowned paper authors**: Usually, a productive or smart paper author may create or contribute to another model. Their GitHub account, X (formerly known as Twitter) account, or other channel updates are good ways to stay aware of their newest work.

- **Using X**: When browsing through my X (formerly known as Twitter) feed, I often come across a mix of content, including humorous videos and images that can easily consume my mornings. To maximize the platform's usefulness, I've learned to utilize the **Not interested in this post** feature to tailor my feed towards displaying more relevant information. This requires discipline, as I need to resist the temptation to engage with entertaining threads and instead interact with AI-related posts that offer valuable insights. By consistently applying this strategy, X transforms into a valuable resource for staying up-to-date with the latest developments in the field.

- **The git pull command**: When we read a post or news from a website, it may be too late to remain relevant or not covered in enough detail. I found that the `git pull` command is a useful way to catch up with the newest advancements in related fields.

  For example, when we open the `diffusers` GitHub repository, and input the `git pull origin main` command, it will list out the latest changes to the main branch, and I can find out what code is merged into the main branch. By simply *ctrl* + clicking on the filenames, I can open the files that include the newest changes.

  Often, the `git pull` command can provide the most valuable information about the cutting-edge advancements of Stable Diffusion. Usually, the code will include the paper URLs too.

- **Useful sites**: Besides GitHub repositories, there are three websites or tools that are very useful for finding papers and models:

  - **Papers with Code** (`https://paperswithcode.com/`): This website is a fantastic resource for staying up to date with the latest research. It provides a comprehensive list of academic papers, along with their associated code implementations, making it easy to understand and reproduce the results. You can filter papers by research area, task, and dataset, and the site also features leaderboards for state-of-the-art models across various tasks.

  - **GitHub Trending**: GitHub is a platform where developers and researchers share their code, and the **Trending** section can be a goldmine for discovering new models and implementations. By filtering the results based on your area of interest (e.g., machine learning, deep learning, or natural language processing), you can find the most popular and recently updated repositories. This can help you stay informed about the latest advancements and best practices in the field.

  - **Hugging Face Model Hub**: Hugging Face is a well-known platform for building and sharing **natural language processing** (**NLP**) models. Their Model Hub is a searchable repository of pre-trained models, which you can filter by task, language, and framework. By exploring the Model Hub, you can find cutting-edge models for a wide range of NLP tasks and easily integrate them into your own projects. Additionally, Hugging Face provides detailed documentation and tutorials, making it an excellent resource for both beginners and experienced practitioners.

While catching up with the newest developments, I found it is also important to stay focused and curious:

- **Maintain focus**: It's crucial for us to stay focused on current tasks or projects without getting too distracted by the latest AI advancements. While it's important to keep up with industry trends, constantly shifting focus can lead to incomplete projects or a lack of depth in understanding. Here are a few strategies to maintain focus:

  - **Prioritize learning**: Identify what you need to learn right now for your project or career goals and focus on that.

  - **Set specific learning goals**: Having clear objectives can help you stay on track.

  - **Allocate time for exploration**: Set aside a specific time each week to explore new advancements. This way, you can satisfy your curiosity without derailing your focus.

- **Stay curious while avoiding being overwhelmed**: It's essential to stay curious and open to new ideas and technologies in AI, but it's equally important to manage this curiosity to avoid feeling overwhelmed. Here are four tips to do so:

  - **Embrace a growth mindset**: Understand that learning is a journey and it's okay not to know everything right now.

  - **Use multiple learning methods**: If one method is overwhelming, try another. For example, if reading research papers is too dense, try watching video lectures or participating in online courses. Alternatively, simply ask a large language model such as ChatGPT for help.

  - **Create a learning schedule**: Plan your learning to ensure you have time to digest new information. Instead of cloning a repo immediately, make a note to schedule a time for it.

  - **Seek support**: Join AI communities, forums, or discussions where you can ask questions and share your learning journey with others. This can make the learning process less daunting.

Balance is key when learning AI. It's about finding the right blend of focus and curiosity that works for you.

Cultivating responsible, ethical, private, and secure AI

As we move forward, AI will become an integral part of our lives, permeating nearly every aspect of our existence, much like electricity and the internet. Initially, these technologies were perceived as novelties, even potentially dangerous. High-voltage alternating current posed a lethal threat, while the internet became a conduit for misinformation. Despite these initial fears, we managed to mitigate the negative aspects of these technologies, harnessing their potential for the greater good.

When groundbreaking technology emerges, it is almost impossible to suppress its spread. Instead of restricting access to these advancements, we should aim to integrate them responsibly into our lives.

AI is no exception. Instances of AI being used for deceptive or fraudulent purposes are not uncommon. As AI becomes more democratized, powerful AI tools will become more accessible to everyone. The challenge lies in managing the potential misuse of AI. Much like a knife, which can be used to both prepare food and cause harm, the impact of AI largely depends on the user.

To prepare for an AI-driven world, we need to increase our understanding of AI, its capabilities, and its limitations. We must learn to use it appropriately and instill ethical and moral values in ourselves and future generations. If possible, we should also establish laws governing its use.

As AI developers, it's crucial to maintain transparency in AI technology. When a large company releases a model that can generate harmful content, such as extreme rhetoric or excessive political correctness, an open community can voice concerns and initiate corrective measures. The same level of AI technology can be used to counteract these issues, essentially fighting fire with fire.

In Stable Diffusion v1.5, a OpenAI CLIP (Contrastive Language-Image Pre-Training)-based safety checker model is available to ensure output is harmless. This model can perform this task automatically and effectively. Stable Diffusion XL also includes a watermarking module, which can embed hard-to-see watermarks in image backgrounds. This feature can protect image authorship by adding specific hidden information. If we keep AI technology open, we can always find ways to balance power and ensure that AI is used for the greater good.

We are at the dawn of a new era, where AI will change the way we live, work, and interact with each other. It's up to us to make sure this transformation brings about positive growth and development, by embracing AI and working together. But there is one thing that keeps haunting us – what if AI takes our jobs away?

Our evolving relationship with AI

On April 24, 1907, lamplighters in New York City went on strike [6], leaving many streets unlit. Despite complaints from citizens and efforts from policemen, few lamps were successfully lit due to various challenges. This event marked a significant shift toward electric streetlights, which were simpler to maintain and had begun to replace gas lamps since their introduction in the late 19th century.

By 1927, electric streetlights had completely taken over, leading to the disappearance of the lamplighters' profession and the Lamplighters Union. The electrification process was unstoppable, no matter how unwilling the public and lamplighters were to adopt it.

And AI is the new electric streetlight; it can be creative, it can be fully automated, it can work on one or several things well, and it may surpass human capabilities. Yes, the AI electric streetlights will replace the old gas lights that we are so used to and carefully maintain. So, will the lamplighters' job be eliminated by the "AI" electric streetlight?

Well, I'm afraid that, this time, it's not so easy to move from gas lamp maintenance to electric lamp maintenance. However, the "AI" electric streetlights are not just replacing jobs but also creating a lot more. And the most important thing is that those jobs created by AI are much more interesting and meaningful than the ones before. To be honest, do we truly enjoy jobs that involve repetitive, mundane tasks, like those once performed by lamplighters? AI will replace boring work and free more of our brain power to explore more interesting, exciting areas that no one has done before. Let's embrace the change, welcome the AI electric streetlights, and start our journey with them together!

Summary

This chapter discussed topics beyond the scope of Stable Diffusion, focusing on the broader context of AI development and its implications for society. Here's a quick summary of the main points:

- The current AI wave is unique because it utilizes attention-based auto-learning architectures, enabling models to transfer knowledge across domains

- Mathematics and programming skills remain essential for AI development, as they form the foundation of AI algorithms and enable researchers to build upon existing knowledge

- You need to stay informed about the latest AI developments through channels such as following paper authors, using X (Twitter), executing GitHub `pull` commands, and visiting useful websites

- Develop responsible, ethical, privacy-protected, and safe AI by promoting transparency, addressing potential misuse, and educating users about AI's capabilities and limitations

- Embrace the transformative power of AI, recognizing that it may replace some jobs but will also create new opportunities for more interesting and meaningful work

By exploring these topics, we can better understand the role of AI in our lives and contribute to its responsible development for the benefit of everyone.

References

1. AlphaGo: https://en.wikipedia.org/wiki/AlphaGo

2. Attention Is All You Need: https://arxiv.org/abs/1706.03762

3. An Image is Worth 16x16 Words: Transformers for Image Recognition at Scale: https://arxiv.org/abs/2010.11929

4. Swin Transformer: Hierarchical Vision Transformer using Shifted Windows: https://arxiv.org/abs/2103.14030

5. Richard Feynman's blackboard at the time of his death: https://digital.archives.caltech.edu/collections/Images/1.10-29/

6. LAMPLIGHTERS QUIT; CITY DARK IN SPOTS; Police Reserves Out in Harlem to Set the Gas Lamps Going. UNION CALLS OUT 400 MEN Only Formed a Short Time Ago, Whereupon the Gas Company Began Dismissals: https://www.nytimes.com/1907/04/25/archives/lamplighters-quit-city-dark-in-spots-police-reserves-out-in-harlem.html

7. Richard Feynman: https://en.wikipedia.org/wiki/Richard_Feynman

Index

packtpub.com

Subscribe to our online digital library for full access to over 7,000 books and videos, as well as industry leading tools to help you plan your personal development and advance your career. For more information, please visit our website.

Why subscribe?

- Spend less time learning and more time coding with practical eBooks and Videos from over 4,000 industry professionals

- Improve your learning with Skill Plans built especially for you

- Get a free eBook or video every month

- Fully searchable for easy access to vital information

- Copy and paste, print, and bookmark content

Did you know that Packt offers eBook versions of every book published, with PDF and ePub files available? You can upgrade to the eBook version at packtpub.com and as a print book customer, you are entitled to a discount on the eBook copy. Get in touch with us at customercare@packtpub.com for more details.

At www.packtpub.com, you can also read a collection of free technical articles, sign up for a range of free newsletters, and receive exclusive discounts and offers on Packt books and eBooks.

Other Books You May Enjoy

If you enjoyed this book, you may be interested in these other books by Packt:

Time Series Analysis with Python Cookbook

Tarek A. Atwan

ISBN: 978-1-80107-554-1

- Understand what makes time series data different from other data
- Apply various imputation and interpolation strategies for missing data
- Implement different models for univariate and multivariate time series
- Use different deep learning libraries such as TensorFlow, Keras, and PyTorch
- Plot interactive time series visualizations using hvPlot
- Explore state-space models and the unobserved components model (UCM)
- Detect anomalies using statistical and machine learning methods
- Forecast complex time series with multiple seasonal patterns

Data-Centric Machine Learning with Python

Jonas Christensen, Nakul Bajaj, Manmohan Gosada

ISBN: 978-1-80461-812-7

- Understand the impact of input data quality compared to model selection and tuning
- Recognize the crucial role of subject-matter experts in effective model development
- Implement data cleaning, labeling, and augmentation best practices
- Explore common synthetic data generation techniques and their applications
- Apply synthetic data generation techniques using common Python packages
- Detect and mitigate bias in a dataset using best-practice techniques
- Understand the importance of reliability, responsibility, and ethical considerations in ML/AI

Packt is searching for authors like you

If you're interested in becoming an author for Packt, please visit `authors.packtpub.com` and apply today. We have worked with thousands of developers and tech professionals, just like you, to help them share their insight with the global tech community. You can make a general application, apply for a specific hot topic that we are recruiting an author for, or submit your own idea.

Share Your Thoughts

Now you've finished *Using Stable Diffusion with Python*, we'd love to hear your thoughts! Scan the QR code below to go straight to the Amazon review page for this book and share your feedback or leave a review on the site that you purchased it from.

`https://packt.link/r/1-835-08637-3`

Your review is important to us and the tech community and will help us make sure we're delivering excellent quality content.

Download a free PDF copy of this book

Thanks for purchasing this book!

Do you like to read on the go but are unable to carry your print books everywhere?

Is your eBook purchase not compatible with the device of your choice?

Don't worry, now with every Packt book you get a DRM-free PDF version of that book at no cost.

Read anywhere, any place, on any device. Search, copy, and paste code from your favorite technical books directly into your application.

The perks don't stop there, you can get exclusive access to discounts, newsletters, and great free content in your inbox daily

Follow these simple steps to get the benefits:

1. Scan the QR code or visit the link below

https://packt.link/free-ebook/9781835086377

2. Submit your proof of purchase
3. That's it! We'll send your free PDF and other benefits to your email directly

www.ingramcontent.com/pod-product-compliance
Lightning Source LLC
Chambersburg PA
CBHW060110090326
40690CB00064B/4722